D0642946

Hands

John Napier

Hands

Revised by Russell H. Tuttle

PRINCETON UNIVERSITY PRESS
PRINCETON, NEW JERSEY

Published by Princeton University Press, 41 William Street
Princeton, New Jersey, 08540
In the United Kingdom: Princeton University Press, Chichester, West Sussex

Library of Congress Cataloging Information—

Napier, John Russell.
Hands / John Napier; edited by Russell H. Tuttle.—Rev. ed.
p. cm.—(Princeton science library)
Originally published: New York: Pantheon Books, 1980.
Includes bibliographical references and index.
ISBN 0-691-02547-9 (pbk.)
1. Hand—Anatomy. 2. Hand—Evolution. 3. Primates—Anatomy. I. Tuttle, Russell. II. Title. III. Series.
QM548.N35 1992
611′.97—dc20 92-14513

First Princeton Paperback printing, with revisions, for the Princeton Science Library, 1993
This book was originally published in 1980 by George Allen & Unwin, Ltd., and is reprinted by arrangement with the Napier estate

Princeton University Press books are printed on acid-free paper and meet the guidelines for permanence and durability of the Committee on Production Guidelines for Book Longevity of the Council on Library Resources

10 9 8 7 6 5 4 3 2 1

Printed in the United States of America

Frontispiece: Albrecht Dürer's "Praying Hands"

Contents

List of Illustrations and Figures

Foreword

PROFESSOR John Russell Napier (1917–1987) is preeminent among the founders of modern primatology. He is renowned for his descriptions and interpretations of the hand of a new species, *Homo habilis* (Leakey, Tobias, and Napier, 1964), and for his creative functional, ecological, behavioral, anmd evolutionary overviews of the order Primates, which stimulated many of his contemporaries and younger students to develop further and challenge his ideas and keen observations. *Hands* is Dr. Napier's last major single-authored work, written at the pinnacle of a distinguished research career. Here he shares his vast knowledge of human and nonhuman anatomy, evolutionary history, and broader anthropological and artistic perspectives on our hands in a highly accessible and entertaining manner.

After perusing the contents of this volume, few, if any, readers will view their hands in the same way or continue to take them for granted. Indeed, don't be surprised if you join the swelling ranks of hand-watchers soon after starting this adventure.

Over the past decade, *Hands* was the only book I could recommend to nonspecialists who wished to know more about the human hand. It is truly an honor to revise this popular classic for a second cohort of general and professional readers.

Above all, I have endeavored to conserve Dr. Napier's voice, style, and wit. He was an exuberant, highly entertaining speaker and raconteur, whose enthusiasm for a topic quickly rippled through an audience. His experiences in medical education, both as student and teacher, and his affable society with persons from many arts and scientific professions supplied him with a rich variety of humorous anecdotes, some of which are to be found in the pages that follow.

In the years since *Hands* was first published, paleoanthropological discoveries and behavioral ecological studies of nonhuman primates have increased dramatically (Lewin, 1989; Tuttle, 1986). In order to preserve Dr. Napier's voice, I did not undertake a major rewriting of the text. Instead, I have annotated the existing text and notably expanded the references so that persons who wish to pursue a particular topic may do so. Moreover, I have provided fuller references on some of the sources that Dr. Napier cited in his text, but did not list in his "suggested reading and references" for the first edition. In a few instances, I deleted clearly inaccurate information vis-à-vis recent research, and I added a paragraph on the hands of *Australopithecus afarensis* under "fossil hands" in Chapter 4. Finally, I substituted currently more socially sensitive words ("humans," "people," "person," and "humankind") for "man" and "mankind," where the referent is to the species *Homo sapiens* or bipedal predecessors that are thought to be related to *Homo sapiens*.

Russell H. Tuttle
Chicago, September 17, 1991

Acknowledgments

ANYONE who writes a book about science or indeed about any subject that is rich in facts, figures, and ideas is from the beginning up to his neck in debt. He must beg, borrow, and steal left, right, and center. His pilfering is usually made respectable by the inclusion of a bibliography, but formal citations do not tell more than a fraction of the story of an author's obligation to colleagues living and dead. Much of one's information is derived from memories of conversations heard, snippets of fact retained, and bright ideas once read, but the source promptly forgotten. To all such anonymous creditors I am grateful and I apologize if their feelings are hurt at finding themselves unacknowledged.

I am happy to take this opportunity of discharging with gratitude some of the debts that have accrued during the writing of this book: to John Barron, John Berry, Audrey Besterman (for most of the diagrams), Michael Bush, Dr. and Mrs. Gardner (for their generosity in allowing me to use a photograph of Washoe never previously published), Ted Grand, Geoffrey Ainsworth Harrison, the late Sir Gerald Kelly, Martin Leadbetter, Gil Manley, Jonathan Musgrave, Prue Napier, Freda Newcombe, Barry Pike (for help with photographs), and Phillip Tobias. Naturally, the opinions expressed in this book do not necessarily reflect the opinions of these experts.

I am grateful to Williams & Wilkins Company, Inc. of Baltimore for permission to quote an extract from *A Method of Anatomy* by J. C. Grant; to Oxford University Press (associated with Geoffrey Cumberlege) for permission to publish extracts from *Elizabethan Acting* by B. L. Joseph, 1951; to Phaidon Press for permission to use diagrams and photographs from my own articles in *Quest for Man*; to Carolina Biological Readers for allowing me to use photographs and drawings from No. 61, *The Human Hand*. I am particularly indebted to the *Illustrated London News* for photographs of the drawings by Neave Parker, the BBC for the photograph of Sir Brian Horrocks, the *Guardian* for that of the Balinese dancer, the Prado for the photograph of the painting of the court dwarf by Velasquez, and to the other museums and individual people whose names appear in the captions to the illustrations that they kindly supplied.

I have also relied heavily on the writings of the late Professor F. Wood Jones, Dr. Sarah Holt, Mr. Michael Barsley, and Dr. Walter Sorell. My apologies if I have failed to credit any author or source.

PART ONE

Nature and Evolution of the Hand

CHAPTER ONE

"You Need Hands . . . "

Hands is a book about the anatomy, function, and evolution of the human hand, but since we are not a one-off product of nature but a species with close nonhuman relatives it is also about the hands of monkeys and apes. Because the functions of the hand touch on so many different aspects of everyday living I have included brief accounts of the role of some of these activities such as tool-making, grooming, and gesture. Such important social issues as handedness and the uniqueness of fingerprints are also discussed.

I hope that *Hands* will appeal to a wide audience of scientists and nonscientists who are, in one way or another, involved in the hand whether as surgeons or dancing teachers, anatomists or fingerprint officers, palmists or geneticists, or just plain citizens who wish to know more about a fascinating subject. With such a wide audience in view, I have tried to write in a light and readable manner, avoiding jargon where possible—or explaining it. Nonscientists may find some parts heavy going, but I hope that this will not impair their understanding. The gap between scientists and nonscientists is not so wide as it used to be, largely because of the influence of radio and television and other forms of science reporting. Equally, I hope that scientists will not be offended by a not-too-technical approach. They may even find it a relief, for, as Margaret Mead put it, "to a physicist even a botanist is a layman." We are all laypersons once we are outside our expertise. There may be only one language of science but there are many dialects.

I can't say that I learned very much about the hand when I was a medical student. I don't believe that many of us did. Certainly it was not a subject of special teaching, and therefore we learned little of its function or comparative anatomy and nothing at all of what might be termed "the aesthetics." Our eyes remained firmly closed to the niceties.

It wasn't until much later, after I was qualified, that I became infatuated with the hand, and embarked on a relationship that has lasted for well over thirty years. Another thirty years and I may begin to understand what the hand is really all about. I remember quite vividly the occasion when the penny first dropped, and I discovered that aesthetically the hand was the most beautiful part of the human body. From then on I began to see all around me the beauty of hands in the context of work and play: in the elegant hands of sculptors, artists, and musicians, as well as in the rough and powerful hands of the village blacksmith.

The hand at rest is beautiful in its tranquillity, but it is infinitely more appealing in the flow of action. The hand of Michelangelo's "David" (Fig. 1) is magnificent but not to my mind so exciting as the hands of Velasquez's court dwarf. David's powerful hand is frozen at rest while the hands of Velasquez's dwarf are frozen in the very act of shuffling a pack of cards. To look at the face as well as the hands is to understand the distinction. David is looking noble but his hand is uninvolved, whereas the hand of the dwarf is very much involved; he is in the process of sliding off the bottom card from the pack. He is no doubt up to one of his conjuring tricks and his sly and arrogant nature is apparent in his evil little face (Fig. 2).

When the hand is at rest, the face is at rest; but a lively hand is the product of a lively mind. The involvement of the hand can be seen in the face, which is in itself a sort of mirror to the mind. One of the saddest sights there is is to watch the hands of the mentally disturbed. When the brain is empty, the hands are still.

My own introduction to the hand came from a young plastic surgeon from New Zealand, John Barron, who was giving an informal talk about the hand to a group of house surgeons at the hospital where we all worked. The gestures he used to emphasize his points were so graphic and so eloquent that I became quite hypnotized and he might have been talking in the Maori language for all that I remember of what he said. I recall going on my late-night ward-round after the talk was over, convinced

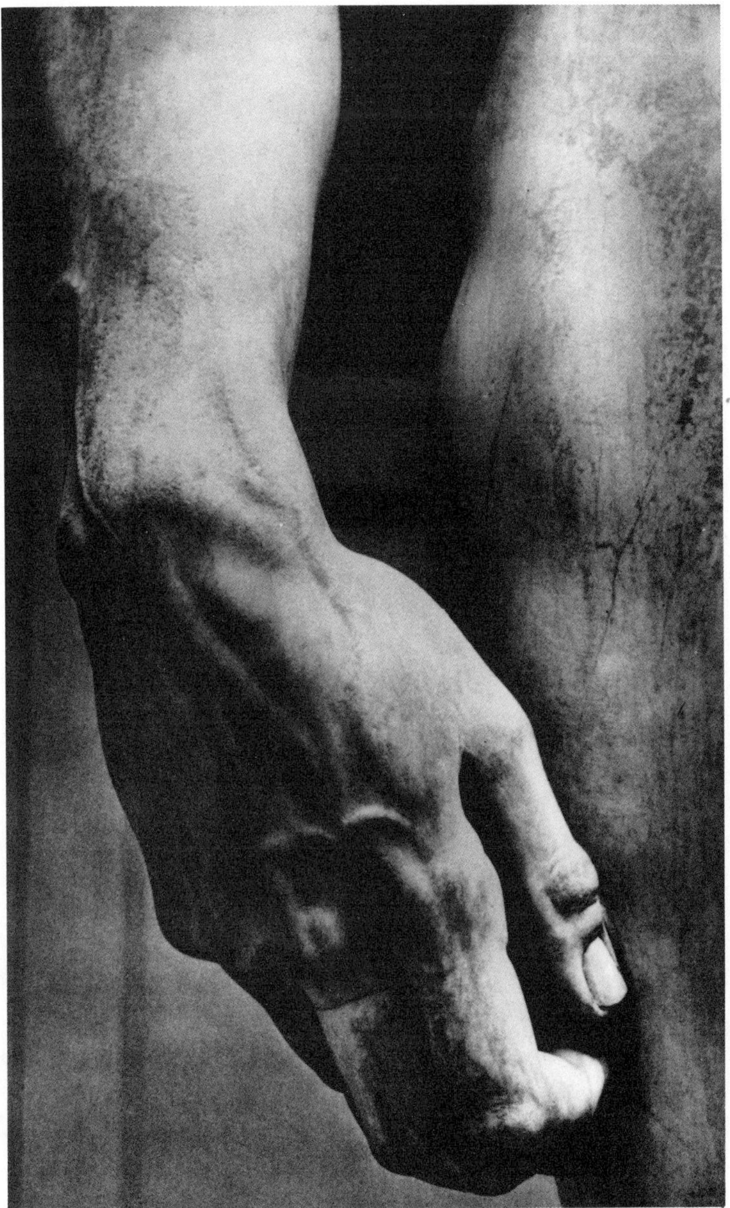

Fig. 1. The powerful hand of David by Michelangelo expresses the position of rest. Compare this figure with Fig. 27.

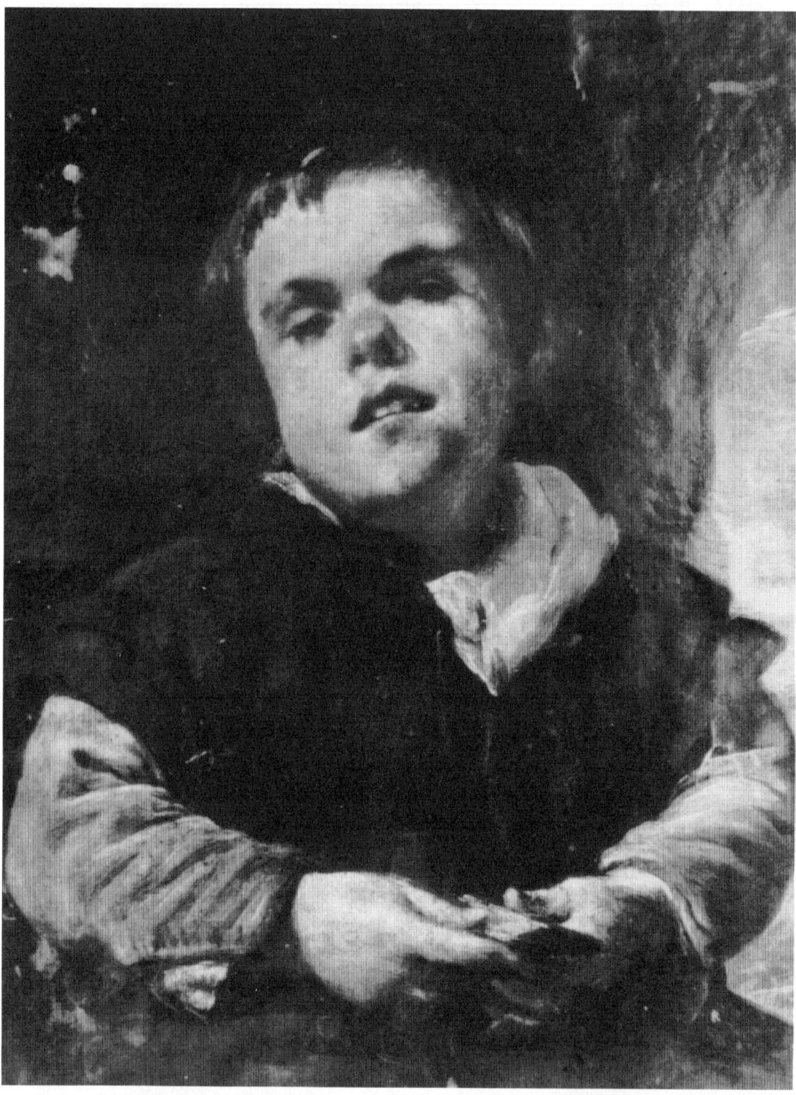

Fig. 2. Velasquez's court dwarf at the time of Philip IV. The facial expression is one of slyness, which leads one to the conclusion that the hands are up to something equally sly. Undoubtedly the arrogant dwarf is performing a conjuring trick. (Courtesy of the Prado, Madrid)

that the hand was going to be my specialization; and in spite of distractions during my career that led me in different directions, I have always returned to the problems of the hand with a feeling of homecoming, so ubiquitous is it in its involvement in human affairs.

My initial interest in the injuries and functions of the human hand has broadened over the years to include its evolution, comparative anatomy, and fossil history as well as the development of tool use in early people. Additionally I have become involved in such side issues as fingerprinting, use of gesture, handedness, cave art, and so on. Most of these intriguing aspects of the subject are included in this book, although not in any great depth.

Strangely enough (and I shall never know whether it was cause or effect), conjuring has always been my principal hobby. No activity demands more of the hands than what is known as "manipulative magic." Unfortunately I was never much good at it, but I learned more about the range and variety of skilled hand movements than at any other time in my life.

All in all, I feel reasonably well qualified to write about some of the aspects of the hand that have interested me. In this sense this book constitutes a personal view and is in no sense an attempt at a definitive study of a very subtle and complicated organ.

There is nothing comparable to the human hand outside nature. We can land astronauts on the moon but, for all our mechanical and electronic wizardry, we cannot reproduce an artificial forefinger that can feel as well as beckon.

But although our hands are such wondrous organs, we tend to take them for granted. Visitors to the zoo indulge in transports of delight at the way an elephant reaches for an apple with its trunk, and become ecstatic at seeing a squirrel use its paws to eat, but give not a moment's thought to the ineffable capabilities of their own hands.

In fact we treat them abominably—with none of the finesse that we lavish on cameras or hi-fi records. This is true of both sexes, but men are particularly prone to mistreat their hands. Fortunately the hand is adapted against such ill-usage by means of thick skin, horny callosities, and a layer of tough fibrous tissue (called the palmar aponeurosis) beneath the skin at the wrist and in the palm of the hand. Even so, infections of the hand are commonplace and injured hands constitute the most common of all industrial injuries.

The human hand, as well as being the principal vehicle of motor activity, is the chief organ of the fifth sense, touch. With the eye, the hand is our main source of contact with the physical environment. The hand has advantages over the eye because it can observe the environment by means of touch, and having observed it, it can immediately proceed to do something about it. The hand has other great advantages over the eye. It can see around corners and it can see in the dark. The hands are situated at the end of long, highly flexible arms, which allows the sensory and motor activities to function at some distance from the body; this form of remote control permits one to feel around corners and so solve one of the great problems facing us in this technological age: it enables one to adjust the knobs at the back of a television set while keeping one's eyes on the screen! Touch as a function of the hand in appreciating three-dimensional art has never been officially encouraged in art galleries; but in 1976 the Tate Gallery in London put on an exhibition of sculpture especially for the blind and encouraged unlimited handling.

How much more telling it would be for children to use their hands to feel the thickness of the fur or the texture of the skin of a mounted exhibit in a natural history museum than to simply stand and stare. The Smithsonian Institution in Washington, D.C., has mounted a large square of elephant skin alongside the main exhibit and they encourage children to handle it as much as they wish.

Dr. Leo Harrison Matthews, in his book *The Life of Mammals*, quotes an example of visual-tactile teamwork. When a friend produces an unusual curio from his pocket you exclaim, "How interesting, let me see it"; but what you often really mean is "Let me feel it."

The hands have one further important function: they are part of our communication system; and in the extent to which they are used to communicate, not only words but also emotions and ideas, our hands are unique in the animal world. Among nonhuman primates, chimpanzees also appear to use their hands to communicate.

One hundred and forty-five years ago the Right Honorable Francis Henry, earl of Bridgewater, left in his will the sum of £ 7000 to the Royal Society. He requested that the money be used for the purpose of sponsoring a number of treatises; one of these famous Bridgewater Treatises—*The Hand*, by Sir Charles Bell—was a very remarkable study in the adaptation of the hand of humans and animals, the more so since Bell was writing at a time when, probably due to the persisting influence of the French naturalist, Buffon, the "imperfections" of animals were a popular device with which to demonstrate the "perfections" of humankind; the sloth, for instance, was much pitied for "its bungled and faulty composition." Those familiar with the absurd anatomy of the sloth will appreciate the mistake, but actually the animal is ideally built for its role in nature. Bell, however, did not fall into this trap; he was clearly aware of John Hunter's principle that "structure was the intimate expression of function" and that function was conditioned by the environment. Lacking the instruction of Darwin (who was at that very time about his business on HMS *Beagle*), Bell was unaware that the catalyst effecting the perfect correlation between structure and function was not a matter of special creation but a matter of natural selection. His thesis was subtitled "its mechanism and vital endowments as evincing design." To Bell, the "design" owed more to divine inspiration at a celestial drawing board than to the trial and error of an earthly workshop. But Sir Charles Bell was a functionalist who appreciated, above all, the relationship between the form of animals and the world in which they lived. In his belief that the mechanism of this relationship was "special creation" he was merely reflecting the scientific convictions of his time.

It is difficult today to realize how great was the impact of Darwin's theory on the fundamentalist way of thought. Darwin's *On the Origin of Species* was received with amusement by some, with profound scientific doubts by others, but with thunderous and righteous wrath by the church, notably in the person of the famous Bishop Wilberforce.

I suppose we all have our heroes. I have three: Hunter, Bell, and Darwin. John Hunter turned our attention from the structure of the hand to its function; Bell related the function of the hand to the environment; and Darwin demonstrated that the environment, by process of natural selection, gave birth to structure. The manner in which an animal responds to its environment is called its behavior; therefore in terms of animals, human and otherwise, we have three levels at which it is necessary to consider any biological problem: structure, function, and behavior—structure, the physical basis; function, the integrated action of the physical parts; behavior, the deployment of function in the setting of the environment. It is at these three levels that I shall be looking at the hand in this book.

The success of an organism ultimately depends on its behavior. In automobile language there is no virtue in a car that doesn't corner or whose petrol consumption is impossibly extravagant. Competition from other cars would soon drive it into obsolescence. In nature the same thing applies to animals. This is in fact the basis of the theory of natural selection. Pursuing the car analogy, natural selection is replaced by consumer choice; no one is going to buy a car that looks and sounds good when ticking over (function) but is incapable for one reason or another of behaving properly in its natural environment—the road.

Human hands show an extraordinary degree of primitiveness—an astounding conclusion when one thinks of their specialized movements, their acute sensitivity, precision, subtlety, and expressiveness. Can the hands of Leonardo da Vinci, a Swiss watchmaker, or a Balinese temple dancer really be called primitive? There is an explanation of this apparent paradox between specialized and primitive. The hand itself is derived from yeoman stock but the factor that places it among the nobles is, as it were, its connections—its connections with the higher centers of the brain. Should the nerve supply be severed by injury or disease, the hand is left high and dry, isolated from its specialized organ of control. The disenfranchised hand is like a beautiful Rolls-Royce that has no engine under the bonnet—elegant but useless. Thus it is possible to argue that while the human nervous system is very specialized, our hands, which retain many ancestral features and have acquired very few new ones, are extremely primitive.

I would like to end this brief introductory chapter with two stories: one is true and the other is probably apocryphal. Both, however, make the valid point that in an emergency the hand is the best surgical instrument ever invented.

The true story concerns myself. As a newly qualified doctor I was confronted one evening with a young girl with appalling injuries to her right leg as a result of a motorcycle accident. When she was brought into the reception bay of the hospital, I could see the main artery of the leg completely severed in the thigh, pumping out blood at a terrifying rate. There were no instruments immediately available and no blood transfusion. So I used my hand as a clamp, pinching off the artery with thumb and forefinger as best I could. Finally I got a bit of string, all that was available, around the artery and tied it off. The blood stopped pumping. It would be nice to report that this unorthodox procedure saved her life, but sadly she died within a few hours. The point of the story is that nothing but the hands could have dealt with that emergency so quickly and effectively. Few patients, thank heaven, ever realize how, during an operation, an appropriately placed finger has saved their lives.

The apocryphal story is about a bluff and arrogant surgeon who was expanding on his favorite topic. "The hand," he said, with his usual pomposity, "is the surgeon's finest instrument." A physician overhearing him, remarked *sotto voce*, "At least the old bugger won't be able to leave *that* behind in the abdomen!"

CHAPTER TWO

Structure of the Hand

I HAVE TRIED to avoid too many technical terms in this book as they make for uncomfortable reading but there are a few that are almost obligatory to any discussion of the hand: the *phalanges* (singular: *phalanx*), *metacarpals* and *carpals*, *thenar* and *hypothenar*, *palm* and *dorsum*, *proximal* and *distal*, and *ulnar* and *radial*.

The *phalanges* are the bones of the fingers, *metacarpals* the bones of the palm, and *carpals* the bones of the wrist (Fig. 3).

Thenar describes the eminence and the muscles comprising it on the thumb side of the hand and *hypothenar* refers to the muscular eminence on the little finger side. *Palm* and *dorsum* (or palmar and dorsal) indicate the front and back of the hand, respectively; they are the equivalents of *anterior* and *posterior*. *Proximal* and *distal* are general anatomical terms that denote the relative position of two structures or regions to each other. Proximal is the nearer of the two to the center of the body in terms of the limb as a whole in the "anatomical position," while distal is farther away. Thus the palm is proximal to the fingers and distal to the wrist. The fingers consist of three bones (phalanges)—the proximal, the middle, and the distal. The *ulnar* and *radial* borders of the hand are named after the two bones of the forearm, the ulna and the radius. When the arm is in the so-called anatomical position in which the palm faces forward, the ulnar border lies on the inner side and the radial border on the outer side.

The idea of the anatomical position underlies all terminology in human anatomy, much to the confusion of zoologists whose subjects do not stand upright on two legs. The concept of an anatomical position is derived from the posture in which human cadavers were in Hunter's time suspended for dissection, either for reasons of space or verisimilitude. In this position the abdomen faces forward and the back faces backward. The palms of the

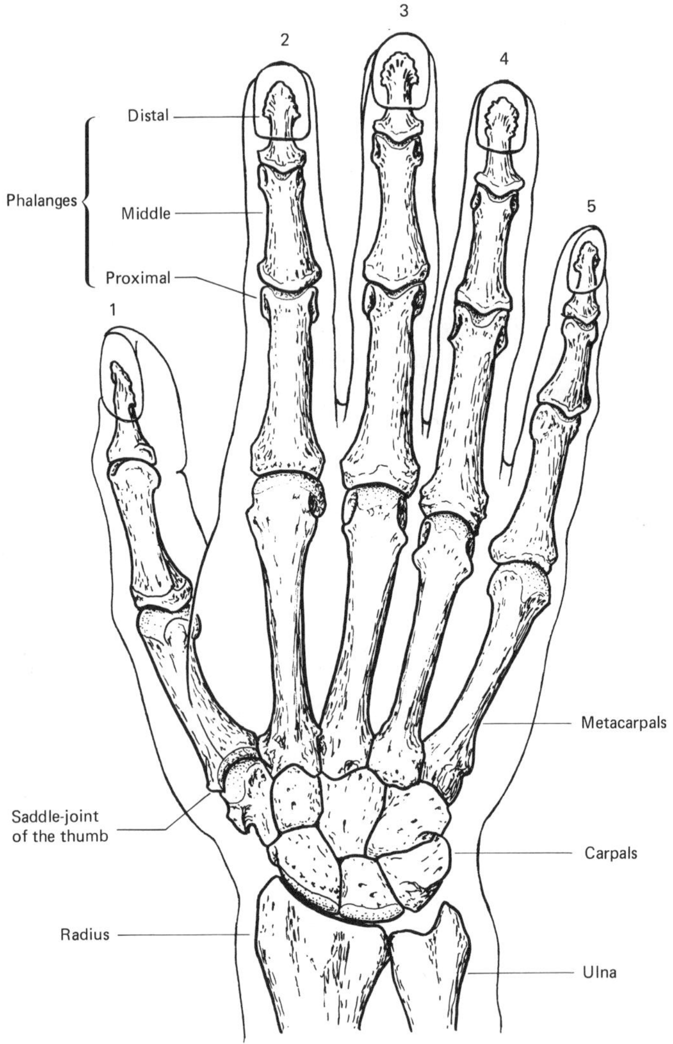

Fig. 3. Bones of the hand seen in relation to the soft tissues.

hands are rotated so they face forward, but the soles of the feet face backward. So, the palm of the hand is said to be directed "anteriorly," but its homologue, the sole of the foot, "posteriorly." Anatomical nomenclature is, to say the least, somewhat confusing. The inward twist of the limbs that takes place during development of the embryo makes the future plantar surface (the sole of the foot) face downward. This occurs in both sets of limbs, but in the upper limb in many species the inward twist is capable of being corrected during life by means of a free movement between the two forearm bones; the hand can be held palm-upward (*supination*) or back-upward (*pronation*). The foot is held in permanent pronation and cannot be corrected. The whole structure of the hind limb in mammals is suborned to the needs of weight-bearing.

The Digits

The Number of Digits

The late Professor F. Wood Jones called the human hand the "absolute bed-rock of mammalian primitiveness." He was referring particularly to the five-fingered state (*pentadactyly*). No mammal, reptile, amphibian or bird has ever evolved a form with more than five digits. There are many that have less, but none that have ever managed to scrape up an extra finger or two as a fixed characteristic of the species. Among the last shots in the war about polydactylism, which was much enjoyed by many mid-Victorian biologists, was a paper by two scientists who in 1892 showed that the multitoed condition in the Dorking fowl was in fact due to an abnormal cleavage of the developing limb bud and was not genetically induced. Strains of six-toed domestic cats, having the same genetic background, are common enough. A rather special case of polydactylism was suspected in the giant panda but Wood Jones (1942) demonstrated with X-rays that the panda's "thumb," which provides such a valuable device for clutching bamboo shoots (Fig. 4), was simply an enormously enlarged wrist bone called the radial sesamoid. The giant panda's thumb is capable of restricted and very simple movements.

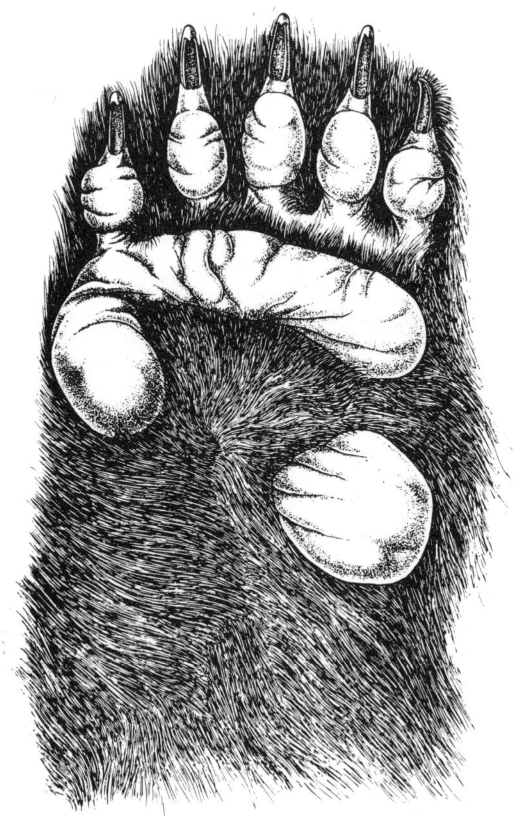

Fig. 4. The false thumb of the giant panda used for clasping bamboo shoots while feeding. (Courtesy of Field Museum of Natural History, Chicago)

Polydactyly is fairly uncommon in people, as is syndactyly, in which adjacent digits are united by webs of skin. The frequency of this congenital abnormality is between one and three per 1000 normal births, with a slight preponderance in favor of males (Fig. 5). Syndactyly, while pathological in people (Fig. 6), is normal for the siamang gibbon, one of the lesser ape family; in fact this characteristic is perpetuated in its scientific name, *Hylobates syndactylus*. The second and third digits of the foot of the siamang are permanently united by a web of skin.

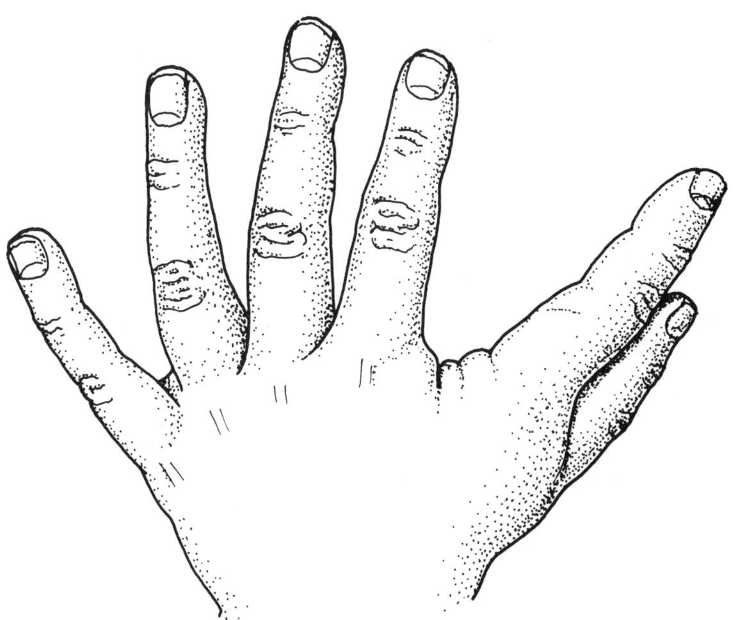

Fig. 5. Polydactyly in a human left hand. (Courtesy of J. N. Barron)

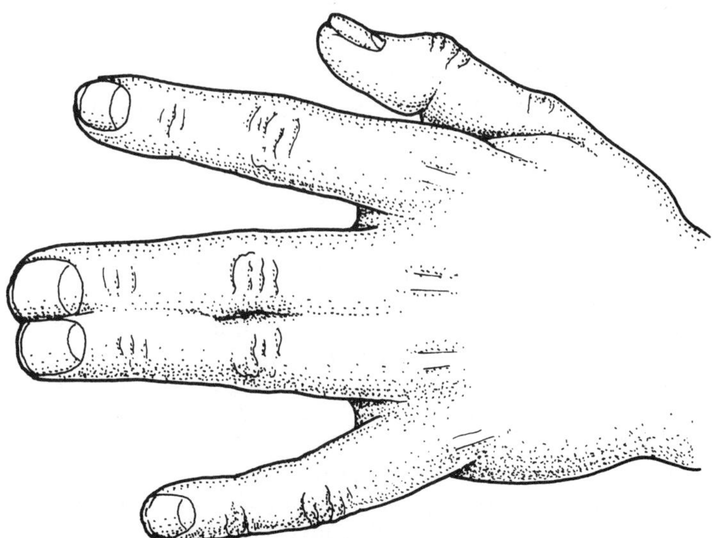

Fig. 6. Syndactyly in a human left hand. (Courtesy of J. N. Barron)

Absence of digits (ectrodactyly) is another example of a situation that is abnormal in people but normal in certain primate species, and indicates how variability provides the raw material for natural selection to work on. What was initially an occasional congenital abnormality initiated during the development of the individual proves advantageous to those individuals possessing it and ultimately becomes a permanent fixture in the genetic code. Three examples of ectrodactyly are the commonplaces of primate anatomy. The spider monkey from the New World and the colobus monkey from the Old are both thumbless, a condition that has nothing to do with zoological affinity of the two species but results from the evolutionary mechanism of parallelism, in which species with similar environmental needs tend to develop comparable physical characteristics. Another example of ectrodactyly is the loss of the index finger in the West African loris, the angwantibo, or golden potto. There is no special virtue in possessing an index finger if you are a potto; indeed there is a distinct disadvantage as it impairs the span of the grasp.

Ectrodactyly is one of the strange conventions of cartoonists for whom the hand consists of only three or four digits. This is true both of human characters and anthropomorphic animal figures. One can only suppose that this foible stems from the simple fact that the hand requires the extremely meticulous draughtsmanship of a Durer or a Watteau if it is not to end up looking like a bunch of bananas. Cartoons essentially are quickly executed line drawings.

The Length of Digits

The term *digital formula* refers to the relative projection of the digits when the hand is laid flat upon the table. The expression was a neologism of Wood Jones, and he explains the occurrence of digits of different lengths in the following lucid paragraph in his book *Arboreal Man*.

> The developing limb of all [vertebrates] appears at first as a small outgrowth from the side of the body. This outgrowth, known as a limb bud, takes the form of a little fleshy paddle with a rounded free edge. It is from the curved free margin that the digits grow outwards

> in a radiating manner. It is obvious if all the digits are of equal length, the middle will project ahead of the others; those standing next on either side will not reach quite so far forwards and the digits upon the extreme edge of the palm or sole will have their tips still further behind.

If the digits continued to develop as they began, the digital formula would be symmetrical ($3 > 4 = 2 > 5 = 1$),* but the digits evolved differential functions in later life and the formula has departed somewhat from the basic pattern of the developing embryo: the thumb has become shorter than the little finger, and the index finger has become longer than the ring finger in a proportion of adults (see Fig. 3).

The third digit (or middle finger) is unequivocally the longest but there is a battle for dominance between the second (index) and the fourth (ring) digits. As it turns out it is a battle of the sexes. A study of 881 college students, predominantly male, revealed that the index was longer than the ring in 28.6 percent. When female records were removed from the series the score was lowered to 22.2 percent, which suggests that females play a large part in keeping the percentage of long index fingers high. When a series of 252 female students was investigated it was seen that the percentage figures shot up to 45 percent, confirming the fact that a long index finger was a characteristic of females.

No such dimorphism exists in primates so far as I know. In my own case, I appear to be bisexual. I have the male formula on my right hand and the female arrangement on my left.

It would be most inadvisable to draw any profound conclusions from the sexual difference in index dominance. Wood Jones, without committing himself in any way, merely labels the phenomenon as another unique and progressive human specialization (zoologically speaking) that emphasizes the independence, or "aloofness" as he calls it, of the human index finger. There is no evidence that relative index finger length has any effect on skill.

* The symbols > and < signify "greater than" and "less than," respectively. The equals sign indicates that the digits are the same length.

Numbers of Bones in the Hand

The retention of a primitive number of bones in the fingers and thumb is hallowed by time. The phalangeal formula of 2 . 3 . 3 . 3 . 3 ., reading from thumb (which has two) to little finger, is the pattern shared by all mammals, living or extinct, and even some mammal-like reptiles, the forerunners of the true mammals. The ancestral amphibian hand exemplified by the fossil *Eryops* displays a phalangeal formula of 2 . 3 . 4 . 5 . 3 ., a pattern that has persisted among reptiles and the majority of mammal-like reptiles. It is interesting to note that the dissident digits are the third and fourth, the leading digits in terrestrial locomotion of all four-limbed creatures. In two-toed ungulates with cloven hoofs the persisting digits are also the third and fourth. It is also notable that the thumb has consisted of only two bones for a very long time. Indeed it is questionable whether it has ever possessed more than two phalanges, a conclusion that will no doubt infuriate the legions of anatomists who collectively have expended much time and energy trying to determine whether the missing element of the thumb is a phalanx or a metacarpal.

The *carpal bones* are eight in number in most living reptiles and in all mammals. The carpals in primates consist of two rows of bones and an intermediately positioned bone called the centrale. From the thumb side the bones of the distal row are called the trapezium, trapezoid, capitate, and hamate; the bones of the proximal row consist of the scaphoid, centrale, lunate, and triquetral. In certain orders, reductions and fusions have taken place; for instance the scaphoid and the lunate of human anatomy are fused in carnivores to form a single element, the scapholunate.

The primate carpal formula is as primitive as it could be, consisting of eight bones arranged in two rows. There are no departures from this theme except in the case of adult humans and African apes, who have no centrale bone; the carpus therefore is reduced to seven bones (Fig. 3). Strictly speaking, the centrale bone isn't missing at all; it is present as a discrete cartilaginous nodule in the very early stages of intrauterine development. Subsequently, it becomes fused with the adjacent scaphoid and is no longer discrete.

Fig. 7. The knuckle-walking posture used by chimpanzees, bonobos, and gorillas on the ground. Friction-skin over the backs of the middle phalanges is tough, but sensitive.

The only distinction between chimpanzees and gorillas and us is the date at which fusion takes place—well before birth in humans and at varying times just before or just after birth in the African apes. Fusion has been accelerated in the case of humans compared with the apes; the actual mechanism is, however, identical.

Wood Jones associates fusion of the human centrale bone with the increasing independence of our thumb and index finger, which requires a stable platform from which to operate; by fusion, one element of the chain is dispensed with. Fusion of the centrale bone in gorillas and chimpanzees has a different history. Both these apes are knuckle-walkers, that is to say, they bear the weight of the front part of the body on the backs of the middle phalanges when walking quadrupedally (Fig. 7). This special posture requires a number of bony adaptations to strengthen the metacarpals, carpals, and wrist joint, including a reduction in the number of mobile bony elements in the carpus.

Terminology of the Digits

The naming of the digits, according to Bertelsen and Capener, can be traced back at least as far as Aethelbert, the first Christian Anglo-Danish king of Kent, who in A.D. 616 laid down a set of laws of compensation for the loss of fingers or thumb. King Alfred and King Canute, both thoughtful—if preoccupied (Alfred) or optimistic (Canute)—sovereigns, revised these laws and in so doing identified each digit by name.

King Canute's rates of compensation for the loss of digits and the scales laid down by the Department of Health and Social Security are a surprisingly close match. They range from 30 percent for the loss of thumb (DHSS) and 30 solidi or shillings (Canute), to 7 percent for the loss of the little finger. Clarkson and Pelly remark that compensation was surprisingly high (18 solidi) for the ring finger compared with 7 percent on the DHSS scale. They suggest the ring finger may have had a religio-mystical significance. There is good reason to believe that the solidus represented a percentage.

The terms used by anatomists in digital identification are fairly obvious (Fig. 8). *Auricularis* (little finger) denotes the digit most commonly employed to extract wax from the depths of the outer ear. The implication of *demonstratorius* (index finger) is self-explanatory, but why *impudicus* for the middle finger? Possibly because it was used as a gesture of derision in Latin and some Arab countries, just as the two-fingered jerking gesture carries a clear insult elsewhere; perhaps because being the longest digit it is ideally disposed to carry out sexual caresses of the female genitals or indelicate scratching operations.

Modern	*Alfred's Anglo-Saxon*	*Canute's Medieval Latin*	*Synonyms*
Thumb-1st Digit	Duma	Pollex	—
Index-2nd Digit	Scythe finger	Demonstratorius	Forefinger, Salutorius, Shooting, Sagittator
Middle-3rd Digit	Midlestafinger	Impudicus	Medius, Famosus, Obscenus
Ring-4th Digit	Goldfinger	Annularis	Medicus
Little-5th Digit	Lytlafinger	Auricularis	Minimus, Pinkie

Fig. 8. The names of the digits.

Professor Wood Jones, a great authority on the hand, preferred the term *obscenus* (middle finger) to describe the digit that is used to express scorn and derision. The ring finger, the *annularis*, is again self-explanatory although its synonym, the once widely used *medicus*, is not; perhaps it is related to the fact that persons formally admitted as doctors of medicine used to wear a gold ring on that finger. One suggestion that has been made is that this digit was used by medieval physicians to stir their cordials and nostrums. Perhaps a latter-day survival of this custom is the trick that generations of consultant physicians have played on their students. The consultant, poker-faced, enquires of the student whether he has tasted the patient's urine. The student naturally admits that he hasn't, whereupon the consultant dips his finger into a specimen jar and touches it to his tongue. What the student, who somewhat unhappily follows his example, is unaware of is that it was the ring finger that the consultant dipped into the urine but the middle finger he put to his mouth.

The intruder (Fig. 8) is the "pinkie," a synonym for the little finger, a term widely used in the United States. Its derivation is obscure, but it possibly derives from a "pink" or "little boat," a word of Scottish origin. The modern terms are the thumb, the index, the middle, the ring, and the little digits; however, there is no reason why "finger" shouldn't be used instead (e.g. index finger). "Ring digit" is distinctly clumsy. Numerical identification using terms such as "second finger" or "first digit" (the thumb) should be avoided as being confusing.

Shape and Proportions of the Hand

It only requires a little observation during periods of ennui while traveling in trains and on buses to conclude that the human hand comes in all sorts of shapes and sizes. The English language is rich in adjectives commonly applied to the appearance of the hand: from beautiful and elegant to powerful and cruel. To describe the texture of the skin we can choose coarse, velvety, smooth, hairy, horny, or moist. To describe its color we have the inevitable lily-white at one extreme and brick-red at the other. To describe function we tend to resort to "professional" analogies: a surgeon's hand, a musician's hand, an artist's hand, a navvy's hand. We also talk of handshakes as being vice-like, flabby, cool, friendly, or warm. Hands are things that people notice in others and on which they place great reliance as indicators of character. They knew (so they say) that so-and-so was not to be trusted immediately when they shook hands with him. Like all generalizations, this sort of typological thinking can lead one into serious errors of judgment. The most brilliant plastic surgeon I know has a short, powerful "square" type of hand that would have been no mean model for the power-orientated fists of our caveman ancestors.

The hands, like the face, are prime areas for the expression of individuality. Individuality is a nontechnical concept that by its very heterogeneity defies classification. The equivalent scientific concept is variability. Variability is the raw material of natural selection; it is what nature selects from. It can express itself in structure and function but, probably more importantly, in behavior. There is an old Yorkshire saying that there is "nowt so queer as folk," which means that one should not be surprised at anything humans do.

Variation in the shape of the hand is particularly obvious. Although function is not unduly dependent on shape, there are undoubtedly some professions that benefit from certain characteristics. Obstetricians are at an advantage in their profession if their hands are long, slender and flexible (the *main d'accoucheur*) as indeed are concert pianists, and magicians whose ability to back-palm half a pack of playing cards depends upon it; they also require a long little finger, which is a variable human feature. Alternatively, the broad, stubby, muscular hand is much more suited to pick and shovel work for which an extensive gripping surface is an advantage.

For all its variation in shape, the overall function of the hand is universal. Some individuals may be "good with their hands" and others "all fingers and thumbs," but the basis for skill doesn't lie in the hand. There are rogue cars that have to have engines replaced within the first few weeks of ownership, but there are no rogue hands. The skill of the hand lies in the brain and it is here that dexterity and adroitness (or clumsiness) originate. The hand is a mirror of the brain; therefore there can be no such combination as dextrous hands and clumsy brains.

I have analyzed the data from many published sources based on measurements of 18,731 individuals (8,480 females and 10,251 males) and it is apparent that the overall *shape index* (the ratio of length to breadth) is remarkably constant (Fig. 9). Furthermore the proportions of the male and female hands differ only very slightly (absolute dimensions apart), the female being on average only a shade more slender than the average male—a rather surprising result. All the data on women stem from the U.S. Army Services and it is possible therefore that there may have been some bias in favor of the robust type of girl with what are known as "capable" hands. Perhaps the Dresden shepherdess-types were weeded out at their medicals. Had data been included from India, Sri Lanka, Burma, and points East, where women's hands are notoriously slender, the results might have been very different.

Among a number of indexes that can be derived from available data on the skeleton of the hand, one, the *phalangeal index*, is particularly valuable in paleontology where there are only bones to go on. It provides insights into the locomotor behavior of fossil apes and monkeys and suggests that the proportions of the human hand have been on the evolutionary production line for at least 20 million years.

Sample	*Number*	*Sex*	*Mean of hand length (mm)*	*Mean of hand breadth (mm)*	*Shape Index* $\frac{\textit{HAND BREADTH} \times 100}{\textit{HAND LENGTH}}$
German Civilians	1200	M	188.0	86.0	45.7
US Army Drivers	2683	M	193.0	88.1	45.6
USAF Flying and Training Personnel	3000	M	190.0	88.0	46.3
Flying Trainees	3000	M	192.0	87.0	45.3
US Army Recruits	8000	F	175.0	77.0	43.9
WAF Trainees	848	F	172.0	77.0	44.8
Total	18,731 Males and Females				Range 43.9–46.3

Sources of Samples: Brezina & Lebzelter, 1953; Daniels & Meyers, 1953; Daniels, Meyers & Worrall, 1953; Herzberg & Daniels, 1953; McFarland & Damon, 1953; Randall & Munro, 1949.

Fig. 9. Shape index of human hands.

The phalangeal index expresses the length of the three phalanges of the middle finger as a ratio of the total hand length. Functionally it reflects in primates the degree to which the hand is adapted for grasping and climbing in an arboreal milieu, or walking and running in a terrestrial one—the rationale being that the longer the fingers the greater is the capacity for grasping; and the shorter the fingers the more stable and powerful is the lever in quadrupedal ground-walkers.

The human phalangeal index has a mean value of 49 percent. This figure is intermediate between that of ground-adapted monkeys such as the baboon, the patas monkey and the gelada (42–44 percent), and the arboreal specialists (52–58 percent). Man shares the 48–50 percent bracket with part-time arboreal Old World monkeys like the macaques, mandrills and the mangabeys which are intermediate in their ecological preferences. With a few exceptions, the lemurs and lorises (collectively known as prosimians) have the longest digits of all, the indicial range showing wide variation with values of 55–64 percent. The long fingers and short metacarpals of arboreal monkeys and prosimians are primitive retentions from the ancestral stock in which long fingers was characteristic.

The hands of great apes are specialized in many ways but particularly in respect of their arm-swinging type of locomotion known as *brachiation*. Although they possess remarkably long fingers, the metacarpal and carpal sections are equally elongated, having kept pace with the fingers during evolutionary growth. Consequently the index value approaches 50 percent and is thus very similar to ours: gorilla 48 percent, chimpanzee 50 percent, orangutan 50 percent.

In practice it is true to say that the living gorilla is almost wholly ground-living. The male seldom climbs trees and the juveniles and females do so only to a limited extent. However, the grounded gorilla does not adopt the typical quadrupedal gait of monkeys but, instead, has opted for a *knuckle-walking* posture (see Fig. 7) whereby contact with the ground is made by the friction pads on the back of the middle phalanges of digits 2–5. Consequently, there have been no evolutionary pressures tending to shorten the fingers because there would be no particular advantage in so doing. The same is true of the chimpanzees, which are slightly more arboreal in habit. Our ancestors never became knuckle-walkers, as did those of gorillas and chimpanzees, but our ape-like ancestors, eschewing a monkey-like quadrupedal gait in favor of bipedalism, also avoided excessive shortening of the hand, the fingers in particular.

An interesting mechanism called *double-locking* occurs in orangutans. Owing to the exceptionally long palm, the tips of the fingers can be tucked into the skin-fold at the base where the finger meets the palm. With further flexion the locked fingers are rolled into the palm (Fig. 10), thus double-locking them. This mechanism is somewhat analogous to an engineer's over-center lock. The adaptive value of double-locking is apparent during feeding. While feeding, orangutans prefer to utilize the slender branches of the trees. They use their feet as well as their hands to distribute the weight evenly between several slim, flexible sup-

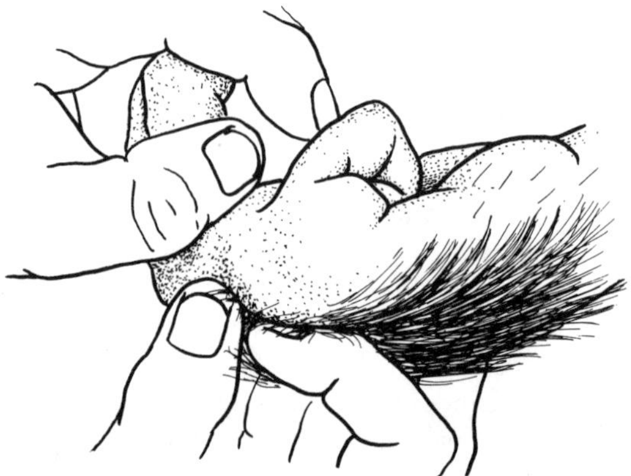

Fig. 10. Double-locking. Orangutans possess long fingers but proportionately longer palms; as a result they are able to tuck the tips of the fingers into the crease at their bases. Further flexion toward the palm constitutes double-locking (see Fig. 11).

ports. It is obvious that for slender branches the most effective type of grip possible is in the double-locked position.

The power of the orangutan's grip by means of double-locking can be demonstrated with a piece of string. People cannot resist a direct pull along the length of the string, but orangutans have no such problem (Fig. 11). The moral here is never take on a tug-of-war team that has orangutans on its side.

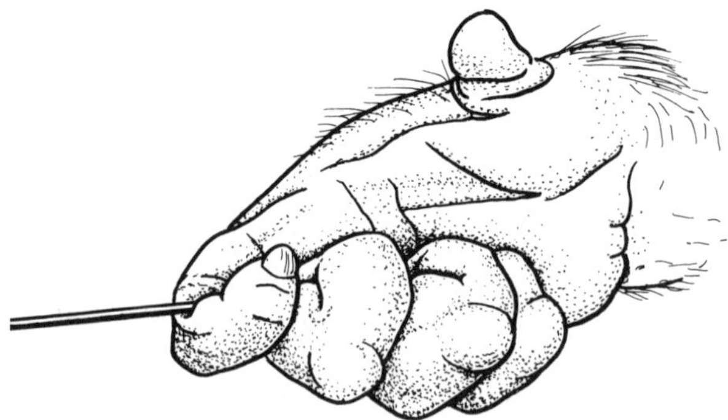

Fig. 11. Double-locking facilitates a powerful grip of slender branches and lianes.

Skin

The skin surface of the human hand, far from being smooth and perfect like alabaster, as so often in romantic novels, is neither smooth nor white but highly textured and multicolored. It is deeply etched with lines, grooves, folds, and furrows and studded with pits for hairs and open ducts for the excretion of sweat. In living hands veins show blue, arteries show pink, and for good measure freckles and "liver spots" show brown; only dead hands are lily-white.

The role of the skin in general terms has been epitomized in somewhat bluff, nontechnical language as the structure that "keeps the blood in, and the rain out." Never mind, the axiom, although crude, provides a starting point. Besides being water- and blood-proof, the skin

> becomes thick where it is subjected to rough treatment, it is fastened down where it is most liable to be pulled off, it has friction ridges where it is most liable to slip. Even with our ingenious modern machinery, we cannot create a tough but highly elastic fabric that will withstand heat and cold, wet and drought, acid and alkali, microbic invasion and the wear and tear of three score years and ten yet effect its own repairs throughout and even present a reasonable protection of pigment against the sun's rays. It is indeed a fine fighting tissue. (S. Whitnall quoted by J. C. B. Grant in *A Method of Anatomy*, 5th Edition, p. 70.)

The skin of the palm is firmly bound to the underlying, packing tissue of the hand. There are areas, over the ball of the thumb and the "heel" of the hand, where the skin is relatively mobile. This is because the underlying fibrous tissue, which is heavily loaded with fat, forms a distinct pad. But elsewhere—as in the central area of the palm—the underlying tissues are fat-free and the skin is firmly adherent.

There was a period during the early months of the last war when among naval gunnery personnel, severe flash burns of the palm of the hand were rather common. They were treated by a full-thickness skin graft (a full-thickness graft includes the fatty subcutaneous layer as well as the skin). The early postoperative results were encouraging. But when the patients started to use their hands again for grasping, the lack of the normal fixation of

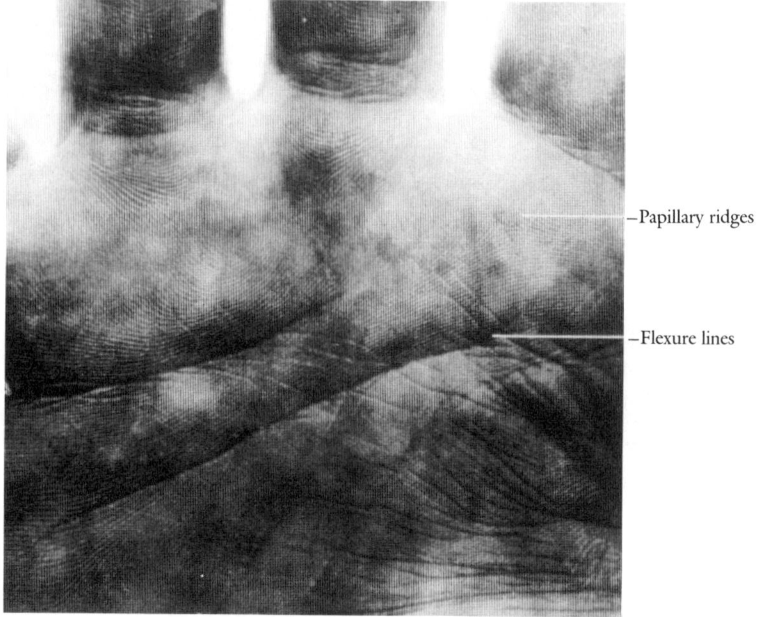

Fig. 12. Palm of hand showing flexure lines and papillary ridges.

fat to fibrous tissue in the center of the palm meant that the grip was slippery and ineffective, and objects thus gripped were flying about like orange pips squeezed between finger and thumb. The lesson was quickly learned and the fatty layer was not included in subsequent skin grafts for this purpose.

Three types of lines can be distinguished in the hand: *flexure lines*, *tension lines* (Langer's lines) and *papillary ridges* (Fig. 12).

Flexure lines are in effect skin hinges that open and close during grasping and gripping, following the movements of the underlying joints. The flexure lines are permanent creases and are analogous with the concertina-like folds that develop in a well-worn jacket across the elbow region. They are lines of skin stasis produced by an anchoring of the skin to the deep packing tissues. Because of their relationship to underlying joints, flexure lines define the principal axes of movements of the hand. However, there is a great deal of variability. There are also a number of subsidiary flexure creases that vary in position in individual hands; subsidiary creases "take up" the excess skin that balloons up when the digits are folded.

The infinite variety of primary and subsidiary flexure lines is a source of great satisfaction to professional palmists whose living is thereby guaranteed. Although there is no scientific support for the idea that palmar creases have any psychic or predictive meaning, it is well known that hands show evidence of occupation, disease, and to some extent temperament and personality—quite an adequate battery of information for the astute palmist to build up a pretty impressive augury. Professor J. D. Lever wrote in 1963:

> In spite of the sceptics [in Roman times], palmistry was considered a worthy subject for study by the intellectuals, and there is the story. . . that Caesar, who prided himself on his infallibility as a palm reader, denounced as an imposter one who appeared before him as Alexander, son of Herod, because his hand did not bear the royal markings. Biblical support for chiromancy can of course be found in the Book of Job (37, 7). . . . He sealeth up the hand of every man that all men he has made may know it.

It is not surprising that with such high level endorsement the ancient art has openly flourished in every age. In our own, palmistry has a somewhat eccentric, although faithful, following.

The most conspicuous horizontal creases in the palm are the *heart line*, which extends from the inner or ulnar border of the hand to the region of the cleft between the index and middle fingers, and the *head line*, which starts in the middle of the palm and curves toward the radial border of the hand paralleling the heart line for a short distance but eventually passing beyond it (Fig. 13). In nonhuman primates the head and heart lines are not separate; instead, they form a single crease that runs straight from ulnar to radial border of the hand without a break. The difference between the simian* type of horizontal crease and the

* Term referring to monkeys and apes.

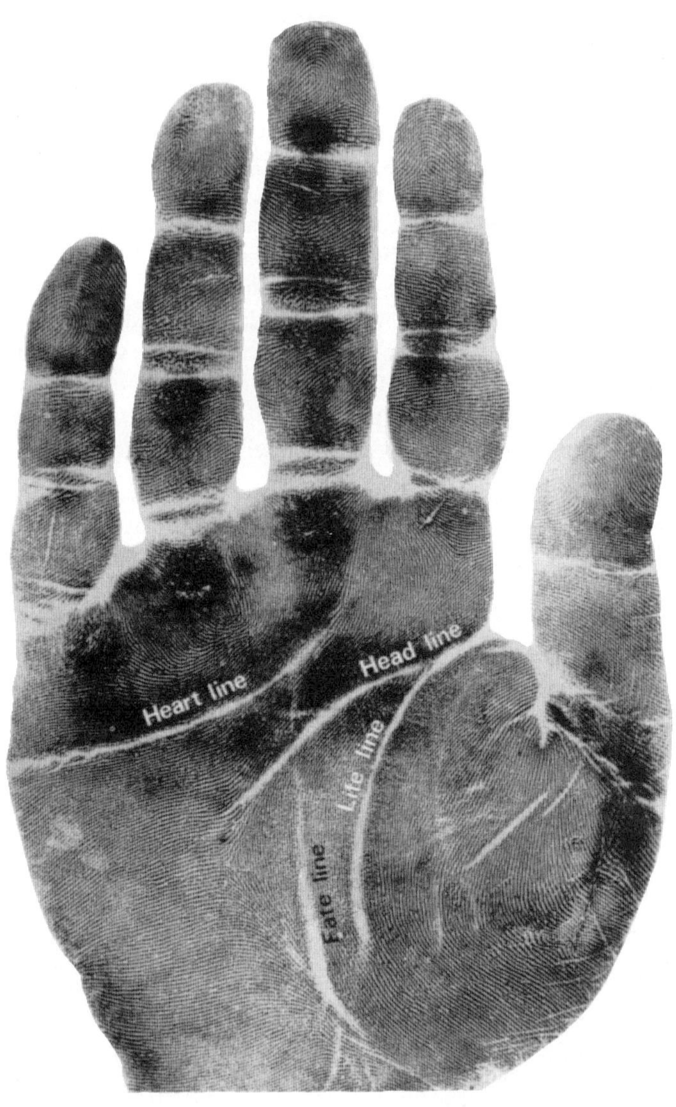

Fig. 13. Hand print showing main flexure lines of palm, named in palmist's terms.

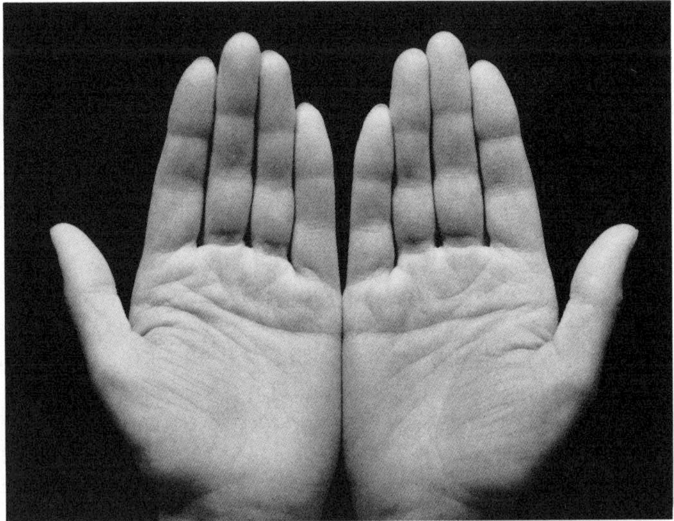

Fig. 14. "Simian crease" apparent on the right; normal on the left. Note 3. 4. 2. 5. formula on both sides. (Courtesy of St. Thomas's Hospital Medical School)

human heart-head lines reflects the fact that the human index finger has a great deal of independent mobility, while, in nonhuman primates, the index finger tends to work in harness with the remainder. A "simian crease" is a fairly rare abnormality in people, cropping up in about 0.4 percent of the population (Fig. 14). Wood Jones summed up the difference between simians and nonsimians in the following terms: "The transverse creases of the human palm are distinguished from the corresponding creases in the hands of monkeys and anthropoids* by their curved form and oblique direction." (Fig. 27).

Among the other flexure lines that constantly appear are those related to the finger joints, which may be double or even triple; it is the distal crease, or—in the case of a triple arrangement—the middle line that most closely marks the line of the joint.

* Outdated term referring to apes. In fact, monkeys, like apes and people, are anthropoid primates because they all belong to the suborder Anthropoidea.

The proximal digital crease is usually single in the little and index fingers and double in the remainder; this difference reflects the extra mobility of the two marginal fingers.

The major palmar creases consist of two consistent horizontal lines (described above) and two or three vertical lines (Fig. 13). The vertical line nearest to the thumb is called the *line of life* by palmists and is the first to develop in the embryo. The second vertical line, which splits the palm into ulnar and radial segments, is the so-called *line of destiny*. These two lines are the principal skin hinge for the thumb.

There are usually two bracelet creases at the wrist. The distal crease, however, which marks the end of the palm, may break up into a double track giving the. appearance of three skin "bracelets." The proximal bracelet corresponds to the wrist joint and the distal bracelet corresponds closely to the transverse axis of the joint between the two rows of carpal bones (mid-carpal joint).

Flexion creases and papillary ridges are especially prone to show abnormalities resulting from growth disturbances in the developing fetus, particularly in cases of genetic abnormality. For instance, in mongolism (now known as Down's syndrome) a "simian crease" is often present and the full complement of flexion creases are absent from the little finger.

Tension lines provide the basis for the wrinkling of the skin that is such a source of anxiety to the middle-aged who still regard themselves as youthful. Tension lines occur all over the body but it isn't until the skin loses its elasticity with advancing years that they form permanent wrinkles. The function of these lines is to provide the skin with a certain amount of "stretchability" in directions corresponding with the natural demands of the region. On the back of the elbow for instance, the tension lines are arranged horizontally, while on the back and on the abdomen they are omnidirectional. Bundles of fibrous tissue interspersed with elastic fibers provide the extensile element over most of the body, but in certain areas, such as the face and the scrotum, muscle fibers are attached directly to the skin. Pronounced wrinkling is usually due to muscle activity. Some wrinkles are congenital, while others, particularly on the face, are acquired (or at least exaggerated) by a lifetime of muscular activity associated with certain facial expressions. "Crows' feet" are a perfect example of tension lines that form with age.

Tension lines in the hand are seen as fine, concertina-like folds arranged horizontally on the dorsum, obliquely on the thenar eminence and vertically on the phalanges. Try pinching up the skin on the back of your hand; either it snaps back into place with the celerity of an elastic band, in which case you are young; or it flattens out sluggishly, in which case middle-age is your problem; or it stays pinched up and you are what is known euphemistically as a senior citizen.

Papillary ridges are permanent thickenings of the epidermis, the outer cellular layer of the skin. They are raised above the general level of the skin, and are found only on the palmar surface of the hand and the sole of the foot. Their distribution corresponds with the principal areas of gripping and weight-bearing, where they serve very much the same function as the treads on an automobile tire. To carry the analogy further, a slight dampness improves the grip of both hand and tire, while wetness leads to a loss of grip in both. Although this is the most popular view, it isn't wholly true that the function of papillary ridges is purely mechanical; their role as outriders for the sense of touch is equally important, perhaps more so.

Ridges vary in coarseness on different parts of the hand. They are at their finest on the fingertips and at their coarsest on the rest of the fingers, the palm of the hand being intermediate. Individually there is a certain amount of variation; there is, for example, a loose correlation between ridge width and body size. A giant has ridges of giant proportions and a midget ridges of midget proportions but the correlation for intermediate statures is not so strong. Professor Harold Cummins has suggested that there are two genetic factors controlling ridge size: the factor controlling body size generally and an independent factor for ridge width.

Papillary ridges are rather like an iceberg: there is as much under the skin as protruding above it. The inner portion consists of a number of finger-like processes intimately related to the organized nerve endings of the epidermis, particularly those that serve touch, which are called Meissner's corpuscles. The papillary ridges act as little lever arms and transmit the most delicate contacts and pressures to the underlying finger-like processes, which are displaced in their turn and so stimulate the Meissner's corpuscles and other endings such as Merkel's disc and free nerve endings.

On the surface, the papillary ridges are arranged in parallel formations, sometimes in curved series, sometimes straight. The visual analogy between papillary ridges and a contour-ploughed field is very close, but the texture of a piece of corduroy with its unidirectional friction effect provides a better mechanical analogy. However, neither of these analogues includes the third and most important feature that complements both the mechanical and sensory functions of the papillary ridges—sweating.

Sweat glands of the common or garden variety, known as *eccrine* glands, are found all over the body in people, but in nonprimate mammals they may be absent from hairy skin and are present only in the hairless walking pads of the extremities. Nonprimate animals have apocrine glands, which are very similar to sweat glands in structure but not in function. Instead of the watery consistency of sweat, apocrine glands produce a thick and viscid fluid that is odorless at first but rapidly becomes rancid as the secretion is decomposed by bacterial action. The typical smell of animals is, at least in part, a function of their apocrine glands. Although body odors are suppressed in modern western society by means of perfumes and aerosol deodorants, people have not always been so prissy. Time was when the village swain would place an apple in his armpit to offer to his partner for immediate consumption at the end of a vigorous barn dance; he clearly was intuitively aware of the strange and potent effect of cheap music and sweaty apples, as Noel Coward might have said.

Among the primates, there is a trend for the eccrine glands to increase at the expense of apocrine glands as we compare prosimians, monkeys, apes, and people. Prosimians have eccrine glands only in the palms of the hands and the soles of the feet;

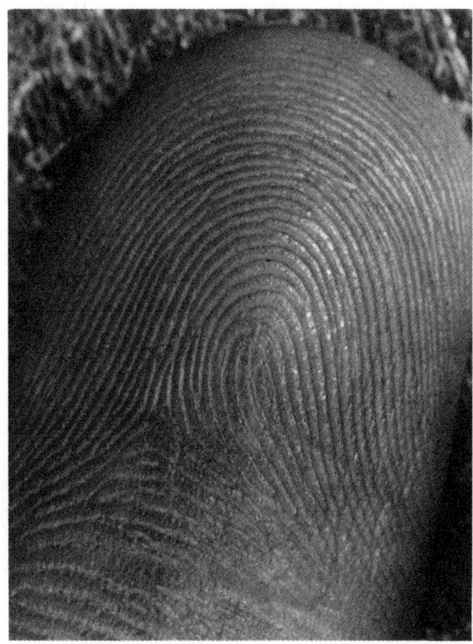

Fig. 15. Tip of finger showing papillary ridges. Note sweat pores along crest of each ridge.

monkeys have a few; chimpanzees and gorillas have more eccrines than apocrines; and people have an enormous preponderance of eccrine glands—amounting to 150 to 340 per square centimeter of skin. We are very sweaty animals.

Eccrine glands lie just below the epidermis, forming coiled cellular tubes; a single duct opens to the surface where it can be easily seen with a hand lens, particularly when it is secreting. In the hand the sweat-duct openings are aligned along the crest of the papillary ridges (Fig. 15). The papillary ridge therefore has a built-in lubricating system, which keeps the skin moist and adhesive and materially enhances the grip of the hand. The dry hand provides a poor gripping surface, as every workman knows who has ever spat on his hands before grasping a pick-handle.

Lubrication of the hands also improves the sense of touch. Both pain and touch are more exquisite in the presence of damp skin. The effect of sweat is that the papillary ridges swell and become more elevated, thus increasing their effectiveness as contact points. By far the greatest concentration of sweat glands occurs in the palm of the hand and the sole of the foot. These glands are particularly interesting. Although sweat glands of the hands are structurally identical with glands elsewhere in the body, they have a slightly different developmental history; furthermore, they cease activity during sleep and they respond not to thermal but to psychic stimuli under stress, a function believed to be mediated by hormones. All three recognized functions of the papillary ridges have at one time been regarded as the most important, but it is a fruitless argument and not worth pursuing. Speaking teleologically of course, you might say the principal function of papillary ridges is to catch criminals.

Hair, Nails and Fat Pads

Desmond Morris has a lot to answer for. No one was too worried about hair and its phylogenetic significance in apes and humans until he put the cat among the pigeons by calling his provocative and important book *The Naked Ape*. It immediately raised a host of questions, one of which was: are people really hairless? There are two ways to answer this: anatomically and physiologically. In spite of appearances, the number of hairs per unit area in chimpanzees and humans is much about the same. The apparent difference lies in the quality of the hair. Human body hair is fine, short, and pale while the chimpanzee's is coarse, long, and dark with melanin pigment. There are slight differences in distribution. For instance, a Caucasian has coarse facial hair, which is lacking in chimpanzees, and chimpanzees have a lower hairline on the forehead, a condition that is frequently met with temporarily in human newborns. Both humans and chimpanzees are prone to baldness at the front of the head. Although male Caucasians appear to be more hirsute than females (Fig. 16), the number of hairs on the body is similar; once again the difference lies in the quality, not the quantity, of the hairs.

Thus, humans are not naked in an anatomical sense, but from the point of view of physiology they are. Since the fine hairs afford no protection against heat loss or the harmful effects of the rays of the sun, they might just as well not be there. While fine pale hairs are useless relics from a hirsute past, the visible

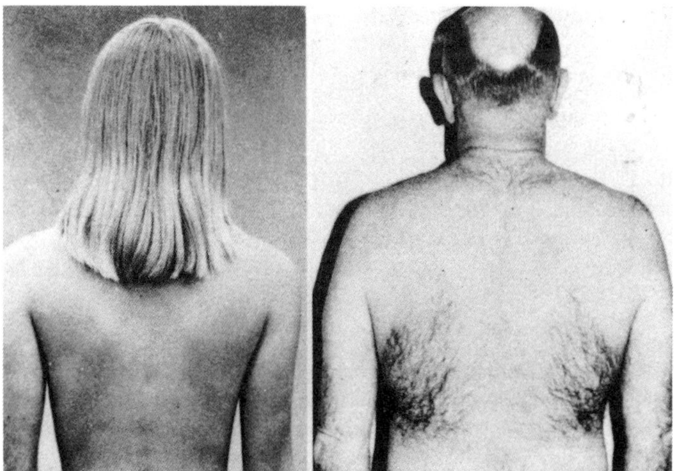

Fig. 16. Male and female Europeans have approximately the same number of body hairs. A misleading difference between them is that the male's body hair is long, dark, and coarse, while the female's is short, fine, and colorless. (Courtesy of Drs. W. Montagna and R. J. Harrison)

hairs of the forearms and legs, for example, function in much the same way as papillary ridges in acting as contact points for the sensory nerves that form a network around the hair follicle.

Although it wouldn't have had such a provocative ring, Desmond Morris might have been more correct to title his book *The Fatty Ape*. One of humankind's great unsung hallmarks is the layer of fat interlaced with fibrous tissue (*panniculus adiposus*) that lies between the skin and the muscles, binding them together and providing for both insulation and fat storage. In humans the distribution of subcutaneous fat is also a secondary sex characteristic (the female breasts, for example). The *panniculus adiposus* is something that we share, surprising as it may seem, with the whales, dolphins, and porpoises, whose subcutaneous fat gives them their sleek, smoothly contoured outline.

The hand is not particularly well endowed with hair; the palm of course is totally hairless and the hair on the dorsum is not very evenly distributed. It is better developed on the inner or ulnar side than on the radial side, and it is present in tufts on the proximal and middle phalanges but it is absent over the knuckles; it is usually absent on the back of the thumb. The hairy coat of mammals (including humans) is not distributed haphazardly; it is arranged in definite tracts according to functional needs. All the hand hair sweeps toward the ulnar border. As Wood Jones points out, if you remember this you need never be at a loss again when your friendly neighborhood police officer requests your assistance in deciding which side an isolated human finger comes from. It comes from the side to which the hairs are pointing.

The central part of the palm of the hand is relatively fat-free, which facilitates gripping; the thenar and hypothenar eminences, however, carry fat pads below the skin.

Nails

With the exception of some prosimians and of marmosets and tamarins—small-sized South American monkeys,—the tips of the manual digits in primates are surmounted by nails. Most mammals that do not have hooves have claws, sometimes sharp and scimitar-like as in cats, sometimes stout and blunted as in dogs. It is generally thought that nails have evolved out of claws opening out to form a relatively flat curved carapace to protect the fingertip. The actual shape of the nail is dictated by the bulk and curvature of the underlying fingertip. In some South American monkeys the tips of the digits are so narrow and the pads so poorly developed that the nails are steeply curved from side to side almost as if they had reverted once more to the more primitive claw.

Claws are not compatible with prehensility because of the mechanical obstruction of claws overgrowing the fingertips; occasionally a marmoset will grip its food between the tips of the claws and the proximal part of the palm but the action lacks any element of precision. Many clawed mammals, especially the smaller ones like squirrels and marmots, hold their food between their two paws as one would a beach-ball; in this way they get around their lack of singlehanded prehensility (Fig. 17).

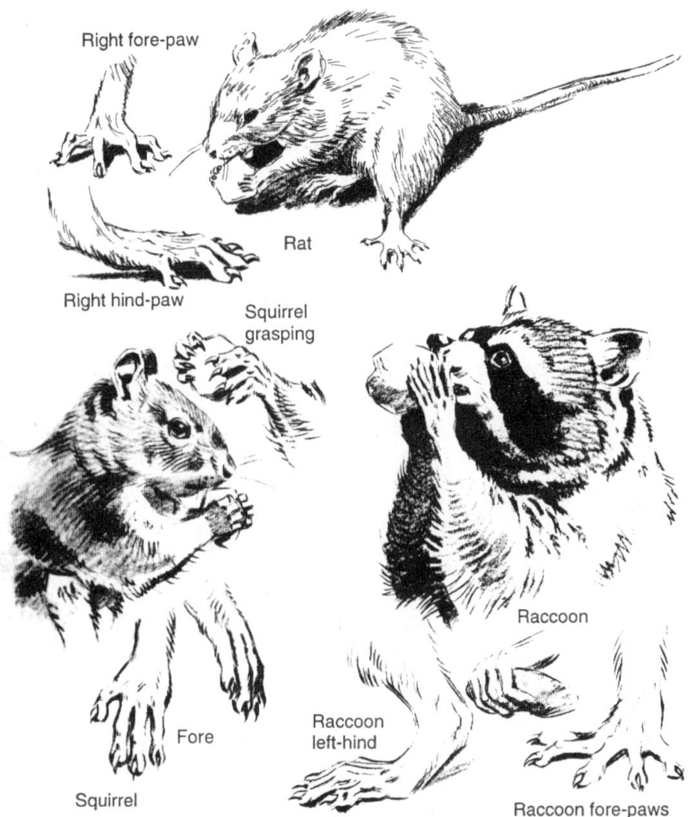

Fig. 17. Two-handed feeding—indicating lack of prehensility—in some mammals. (Courtesy of the *Illustrated London News*)

Human nails are composed of keratin in the form of flat, compacted cells derived from the skin. They develop in the human embryo between the eighth and twelfth week of fetal life. Nails themselves are dead tissue but are set on a plate of living flesh rich in blood vessels and nerve endings. The source of new growth is a matrix of germinal cells buried beneath the fold of skin at the base of the nail. As they push forward the cells from the matrix become keratinized.

The growth of nails is continuous. The speed of growth depends on a number of factors such as age and the state of health. Nails are said to grow faster on the side of the dominant hand; nail-biting, too, is considered to be a stimulating factor. The ambient temperature is also believed to have an effect on growth rates, being faster in summer than in winter. The thumb is the fastest growing nail and the little finger the slowest. To some extent the size of the half-moon at the base of the nail is a guide to the speed of growth—the larger the faster. All things considered, the overall growth of human nails is of the order of 0.1 mm per day.

The responsiveness of nails to a variety of conditions, in the internal as well as the external environment, has interested physicians over the years, and nails constitute an important, if unobtrusive, element of any doctor's examination. Nails are the barometers of health. Longitudinal ridges, brittle nails, transverse grooving, the size of the half-moons, the color of the nail-bed seen through the translucent nail, and white marks in the keratin are all indicators of health in one way or another. Any severe illness (physical or psychiatric) may leave transverse linear depressions on each nail. These are called Beau's lines.

The principal function of nails is to provide both a rigid backing and a protective carapace for the pulpy fingertip, which is of such critical importance for the manipulation of small objects and the discrimination of textures. The secondary functions are legion—from winding a watch to opening a cellophane-wrapped package. Habitual nail-biting may have many sinister temperamental overtones, but a far worse deprivation is the loss to the nail-biter of a built-in tool-kit of cutters, pliers, scrapers, and screwdrivers.

Palmar Pads

The primitive mammalian "hand," which is, of course, functionally used as a foot, bears a maximum of eleven pads. There are five apical pads (one at each fingertip), four interdigital pads in the distal area of the palm, and a thenar and a hypothenar pad in the proximal palm. The hypothenar pad is often divided into

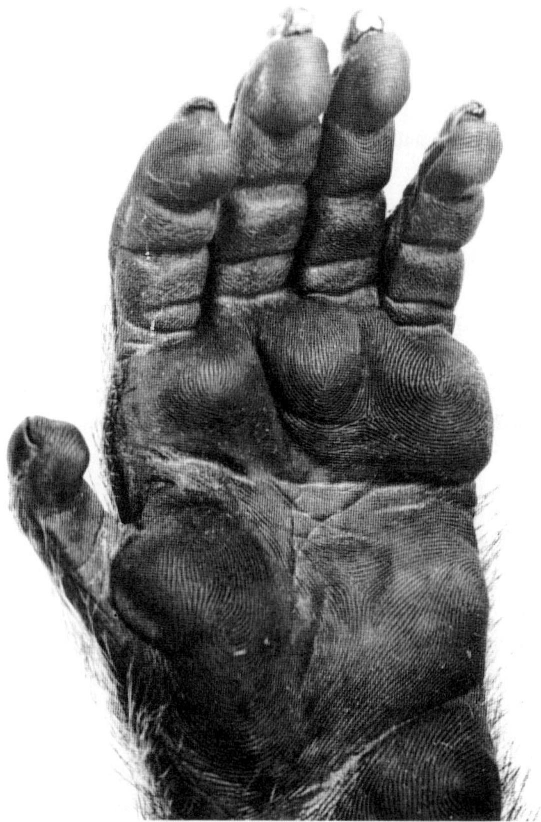

Fig. 18. Palmar surface of the hand of an Old World monkey. Note palmar pads surmounted by papillary systems.

two parts (Fig. 18). Each palmar pad is surmounted by skin that bears papillary ridges. This is true even of people, whose hands still bear the traces of the primitive quadrupedal walking pads. We retain the five apical pads on the ends of the fingers, three interdigital pads, and a hypothenar pad. The thenar pad has disappeared, the pad or "mount" of the thumb being largely a muscular eminence. The interdigital pads are barely discernible.

The remnants can be seen as small puffy swellings at the base of the clefts between the fingers. The hypothenar pad serves to cushion the pressure exerted by the handles of tools and weapons held in a power grip (see Chapter 4). Below the skin of the hypothenar pad and partly attached to it is a small muscle called the palmaris brevis whose function is very mysterious. It helps to "cup" the hand forming the "Cup of Diogenes" (named after the Greek philosopher noted for his simple mode of life) and serves to stiffen the hypothenar fat pad, thus giving it more rigidity for power-gripping purposes. As might be expected, there is no comparable muscle in the hands of monkeys or apes; it would appear to be a bona-fide hallmark of humankind.

The hypothenar pad is homologous with the heel pad of the human foot. In fact this part of the human hand is often referred to as the "heel." Like all palmar pads, the heel pad is made up of lobes of fat encapsulated in a network of fibrous tissue. This type of adipose tissue is not merely a depository for fat like the abdominal wall, but a structure with an essential functional role, that of defending underlying tissues from injury.

When the starving internees of German concentration camps were finally freed in 1945, most of them had used up all their available stored fat, even from the walls of the heart. Yet in spite of the fact that the prisoners were little more than walking skeletons, their heel pads were intact, rounded, and as fatty as ever, and (although I don't recall any precise mention of this) I suspect that their hypothenar fat pads also survived.

Internal Anatomy

With the glove of skin stripped from the hand and the internal structure revealed, one begins to understand what is meant by *multum in parvo*. In an incredibly small space, tendons, muscles, nerves, veins, and arteries are packed as close as sardines in a tin. It is possible to wax quite lyrical over the functional effectiveness of the anatomical arrangements of the hand as Sir Charles Bell did: "and we must confess that it is in the human hand that we have the consummation of all perfection as an instrument."

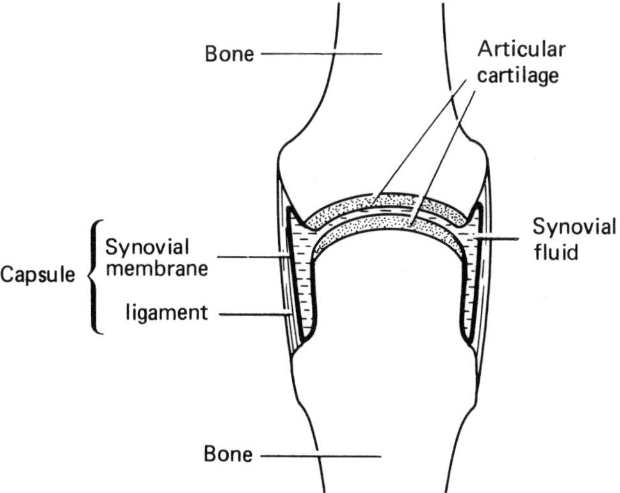

Fig. 19. A hypothetical synovial type of joint found in humans and all other mammals.

Joints

The skeleton of the hand is comprised of twenty-seven elements, of which fourteen are phalanges or finger bones, five are metacarpals or palm bones, and eight are carpals or wrist bones (see Fig. 3). All of them articulate with at least one other bone by means of a synovial joint. Briefly, the synovial joint (whose essential feature is its mobility) is the most common joint in the body. Movements are controlled by muscles and ligaments; if there are four movements possible at any given joint then there are at least four sets of muscles, one for each. Without muscles, the limbs would flail—as uncontrollable as a marionette without strings. Synovial joints are enclosed in a capsule lined by a moist, glistening synovial membrane made up of cells that secrete the lubricating fluid that makes for the smooth action of a joint as well as supplying nourishment. The outside of the capsule is strengthened by ligaments that also act to some extent to guide the movements of the joint (Fig. 19).

There are several sorts of synovial joints, which are classified according to the type of movements permitted. Joints with a single axis are of two kinds—the hinge and the pivot. The hinge-type operates like the lid of a box that opens and shuts moving about a single axis (the elbow is a good example). In the pivot joint, movement is limited to one of rotation around a fixed point (the skull and atlas vertebra pivot on the axis vertebra to allow twisting movements of the head on the neck). Bi-axial joints, or condyloid joints, are those that have two axes of movement at right angles to each other; poly-axial joints have multiple axes and they operate on the principle of a ball-and-socket. There is also rather a special joint—of immense significance to the working of the hand—called a saddle joint, which combines the properties of a bi-axial joint and a ball-and-socket joint. Finally there is the so-called plane joint, which allows only a sliding movement between the two opposed surfaces. In spite of its name, the surfaces of a plane joint are often slightly curved; the action of a plane joint can be mimicked by the movement of two stacked saucers sliding on one another.

All synovial joint types, with the exception of the pivot and the ball-and-socket, are represented in the hand: hinge joint—phalanges articulating with each other; bi-axial joint—phalanges articulating with metacarpals; saddle joint—the thumb metacarpal articulating with the carpal bone (the trapezium); plane joint—joints between adjacent carpal bones.

Movements of the Fingers

The fingers can be bent (flexed) or straightened (extended) at the hinge joints between the phalanges. In addition, the condyloid joint between the proximal phalanges and metacarpals allows movement from side to side and at right angles to the plane of flexion and extension. Movements away from an imaginary line drawn through the middle of the middle finger are termed *abduction* and toward it *adduction*. As in all condyloid joints, a very small degree of rotation can also take place; this movement is particularly noticeable in the index finger and forms an important element of opposition movements (see below). The muscles that produce movement at the metacarpophalangeal joints are the flexors and extensors of the fingers that have their proximal attachment in the region of the elbow, aided by the small muscles arising within the hand called the *lumbricals* and *interossei*.

There are two systems of flexor muscles (Figs. 20, 21), the superficial and the deep. Passing from the forearm to the hand, the muscular portion of the muscle is replaced by a tendon, which has the effect of reducing bulk. The tendons forming a circumscribed bunch pass through a narrow tunnel consisting of the carpal bones bridged over by fibrous tissue, and spread out fan-wise so that one deep and one superficial tendon passes into each finger, the deep being behind the superficial.

The deep tendon acts primarily on the last joint of the fingers, while the superficial tendon is responsible for flexing the middle joint. The method by which this seemingly impossible trick is accomplished is beautiful in its simplicity. The tendon of the superficial flexor splits into two, each half passing round the deep tendon with a twist, rejoining behind it and passing distally in this new relationship until it is inserted on the middle phalanx.

The long flexors of the fingers are also adductors, in the sense that not only do they bend the digits but they also produce adduction when the tendons contract. You will notice that on bending the outstretched fingers the tips approach one another as they fold into the palm. In fact, they begin to get in each other's way. The reverse happens (abduction) when the hand opens out. Adduction of the digits (including the thumb) provides the principal mechanism of grasping in nonprimate mammals (Fig. 17), and substitutes in some measure for true prehensility. Although primates have evolved a more sophisticated form of grasping using the opposable thumb, adduction (or convergence) on bending and abduction (or divergence) on straightening are movements still retained by the human hand.

An important adjunct to grasping is the extension of the wrist-joint. The wrist in 30-40 degrees of extension is in the optimal position for gripping, whether this is mass flexion as in gripping a hammer, or individual flexion as in tapping typewri-

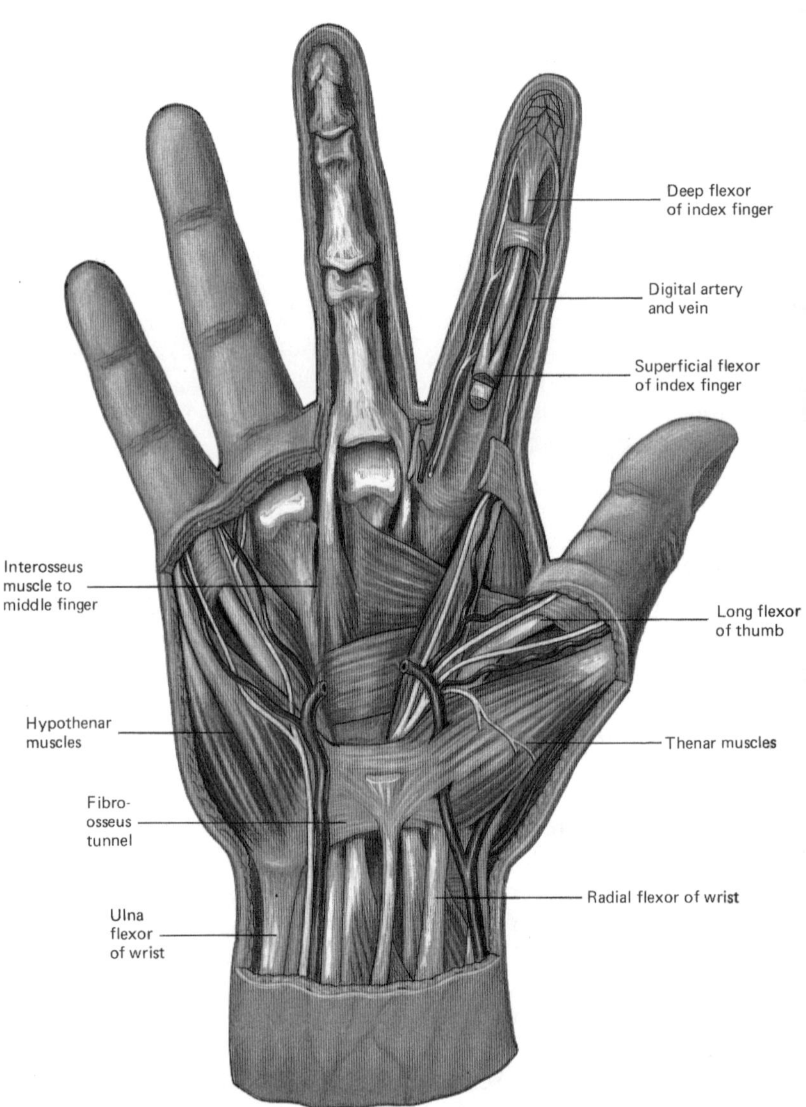

Fig. 20. Palmar view of a dissection of the hand.

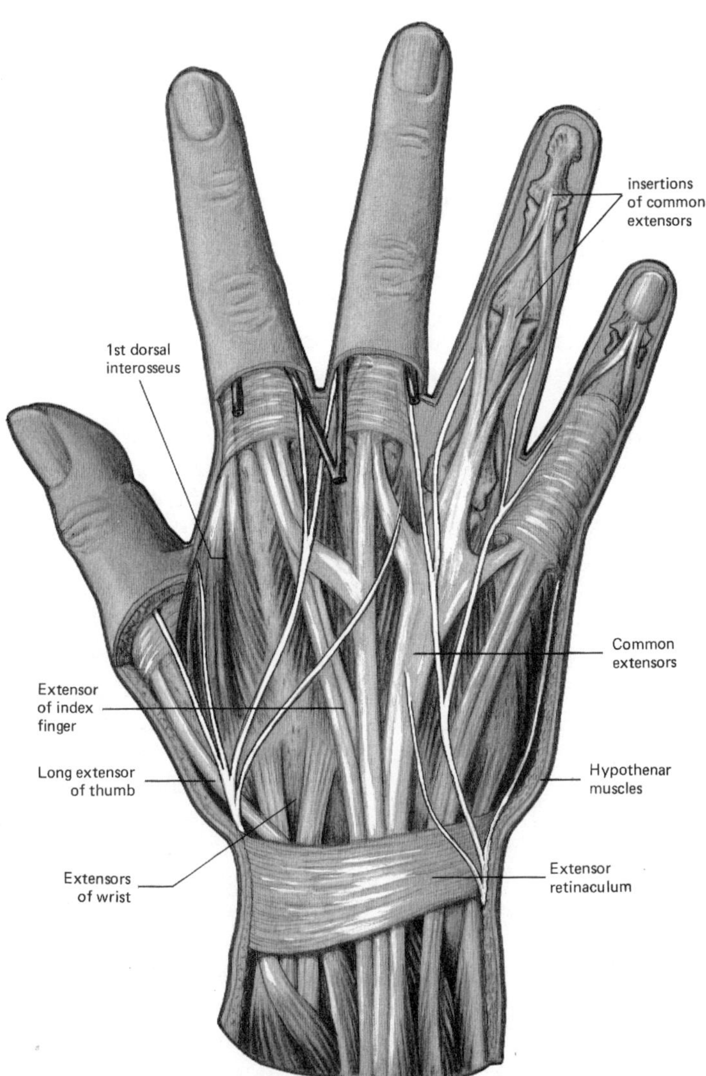

insertions
of common
extensors
1st dorsal
interosseus
Common
extensors
Extensor
of index
finger
Long extensor
of thumb
Hypothenar
muscles
Extensors
of wrist
Extensor
retinaculum

Fig. 21. Dorsal view.

ter keys. In full flexion of the wrist, the hand can exert only 25 percent of its power in full extension. As is well known, the quickest way to disarm an assailant wielding a knife or a gun is to force his wrist into flexion, whereupon he will be forced to drop his weapon, or at least that's the principle!

The long extensor muscles of the fingers differ somewhat from the long flexors at their insertion; to begin with they are not arranged in two layers. Each tendon spreads out into a flattened sheet—what is known as the extensor assembly—over the backs of the fingers (instead of remaining as a rounded tendon, as is the case with the flexors). The human index finger has two extensors, a superficial and a deep; the rest have only one.

Most Old World monkeys such as the macaques have a full set of extra extensors, one for each finger. The orangutan has a nearly full set of "extra" tendons but gorillas and chimpanzees, like humans, possess only an "extra" tendon for the index finger. The index finger is particularly adept in the gorilla, having an almost human-like "aloofness."

Each tendon of the long extensor divides into three slips (Fig. 21): the central slip becomes attached to the proximal phalanx; the other two stream on as a rather undistinguished sheet of fibrous tissue that, once again, forms three slips. The central slip reaches the middle phalanx where it is attached, and the outer two reach the distal phalanx. The tendons of the short muscles of the hand (the interosseus and lumbrical muscles) join the assembly and play an extremely complex role in the mechanism of the hand. These short muscles are extremely well endowed with special nerve endings that provide them with a positional sense that has no equal elsewhere in the body.

Acting independently of the extensor assembly, the lumbricals produce radial deviation of the digits. A paralysis of the lumbrical muscles that occurs following injury to the nerves reaching the hand (see below) accentuates the normal tendency of the fingers toward ulnar deviation. Extreme ulnar deviation—or "ulnar drift" as it is called by clinicians—is one of the most striking features of rheumatoid arthritis (Fig. 22). The deformity comes about as a mechanical result of laxity of the capsules

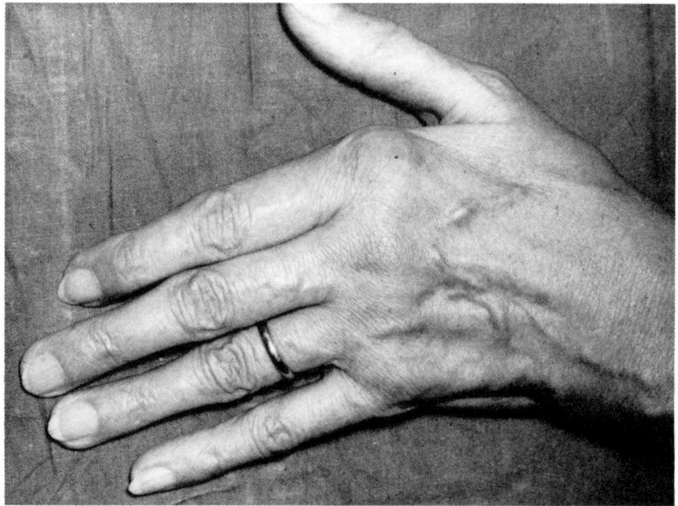

Fig. 22. Ulnar drift, resulting from imbalance of intrinsic hand muscles following inflammation. A common deformity in elderly people with rheumatoid arthritis. (Courtesy J. N. Barron)

of the knuckle joints due to inflammatory condition of these joints, together with a loss of function of the interosseus and lumbrical muscles for the same reason. Dislocation ulnarward of the extensor tendons into the gutter between the knuckles follows; the natural tension of the extensor tendons now simply accentuates the normal tendency toward ulnar drift.

Movements of the Thumb

The thumb, the "lesser hand" as Albinus called it, is the most specialized of the digits. Isaac Newton once remarked that, in the absence of any other proof, the thumb alone would convince him of God's existence.

The generalized mammalian thumb is not a particularly specialized digit and if it has any unique qualities, in a sense they are attributable to its marginal position. Being unsplinted by adja-

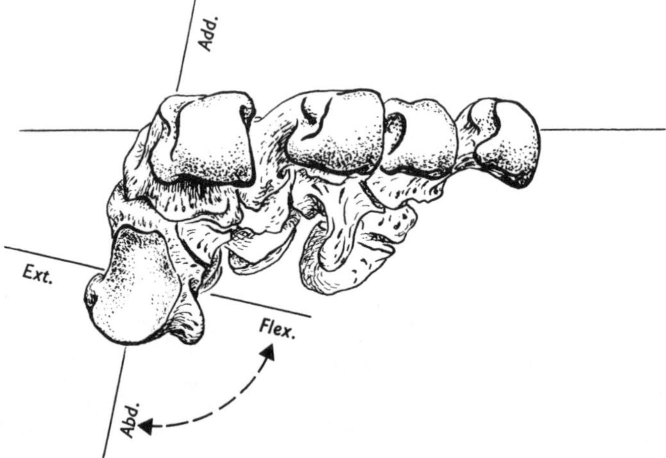

Fig. 23. Metacarpals 2–5 seen end on. The first metacarpal has been removed to show the trapezium. Axes of movement of the thumb are indicated. A combination of abduction and flexion results in opposition. The saddle joint allows a rotational movement (indicated by the broken line).

cent digits, it is therefore free to develop a minor degree of extra mobility; the same is true of the little finger.

The Primates (with a very few exceptions) show some degree of functional independence of the thumb, although the extent and precision of its movements differ considerably among the members of the order. One can see a steady increase in the functional complexity of living primates from the prosimians (lemurs and lorises) to the nonhuman anthropoid primates (monkeys and apes) and finally to humans. Monkeys are more adept than lemurs, apes are more adept than most of the monkeys, and people are more proficient still.

The thumb metacarpal is unique. Alone among the metacarpals, it articulates with the carpals by means of a freely movable saddle joint. The remaining carpals are of the plane joint variety, which have very small ranges of movement. The carpometacarpal joint of the thumb, being of the saddle type, is almost as mobile as a ball-and-socket joint and has the following movements: adduction-abduction, flexion-extension, and medial-lateral rotation (see Fig. 23).

The functional advantage of a saddle joint is that the two opposing surfaces and their supporting ligaments are so arranged that the stability of the joint is provided without the need for a cuff of bulky muscles disposed around the joint to control and direct its movements, as is the case with other ball-and-socket joints like the shoulder and hip. Bulky muscles at the root of the thumb would seriously impair its manipulative skill and flexibility.

CHAPTER THREE

Function of the Hand

Opposition

Perhaps the most important movement of the human hand is opposition. The movement of the thumb underlies all the skilled procedures of which the hand is capable. The hand without a thumb is at worst, nothing but an animated fish-slice, and at best a pair of forceps whose points don't meet properly. Without the thumb, the hand is put back 60 million years in evolutionary terms to a stage when the thumb had no independent movement and was just another digit. One cannot emphasize enough the importance of finger-thumb opposition for human emergence from a relatively undistinguished primate background. Through natural selection, it promoted the adoption of the upright posture and bipedal walking, tool-using and tool-making that, in turn, led to enlargement of the brain through a positive feed-back mechanism. In this sense it was probably the single most crucial adaptation in our evolutionary history.

As it is preferable that opposition should describe the movement of the thumb as a whole, the relative contributions of the three different joints involved are ignored in the following definition:

> Opposition is a movement by which the pulp surface of the thumb is placed squarely in contact with—or diametrically opposite to—the terminal pads of one or all of the remaining digits (Fig. 24).

From this definition it is clear that all the digits are involved in the concept of opposition, not merely the thumb. Because it is the most important finger, as well as for the sake of descriptive simplicity, the index is used as the opposing digit. Finger-thumb opposition hereafter means index finger-thumb, although much the same arguments apply equally to the other fingers.

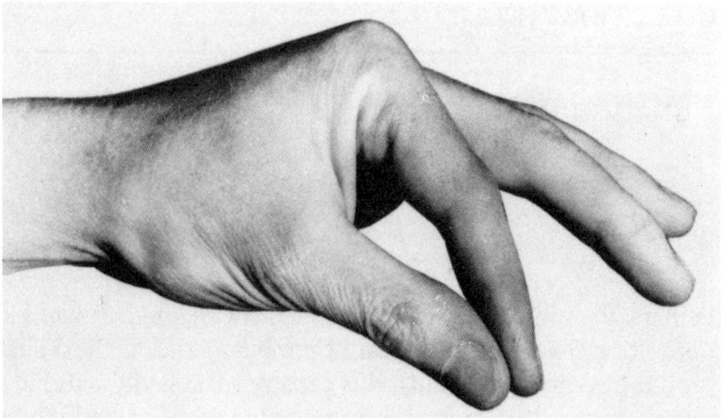

Fig. 24. "Perfect" opposition between thumb and index. Note broad area of contact and slightly extended wrist joint.

The neutral position of the thumb (analogous to a gear-lever "in neutral") is taken as the starting point for a description of the movement of opposition. When the thumb is abducted, it angles from the palm, as the muscles spring into action—into gear, as it were. Once in gear, it moves ulnarward with a movement of flexion and rotation; the more it flexes, the more it rotates until it comes to a full stop in full opposition. To complete the movement, the index finger flexes at all its joints until it makes contact with the tip of the thumb. Because the thumb is rotated, and because the index finger contributes its own element of rotation, an intimate contact between the digital pads of the finger and thumb is achieved (Fig. 24). The role of the index finger is relatively passive. Apart from the long flexors, the most important muscle acting on the index finger is the dorsal interosseus, which lies between the thumb and the index finger. It is easy enough to see and feel the interosseus in action when the opposing finger and thumb are tightly pinched together and then relaxed.

Although opposition of the thumb is one of the hallmarks by which humans can be authenticated as it were, it is not a movement unique to us. What is unique is the broad area of intimate contact between the fingertip pulps of the opposing digits that results. The advantage of intimate contact concerns both func-

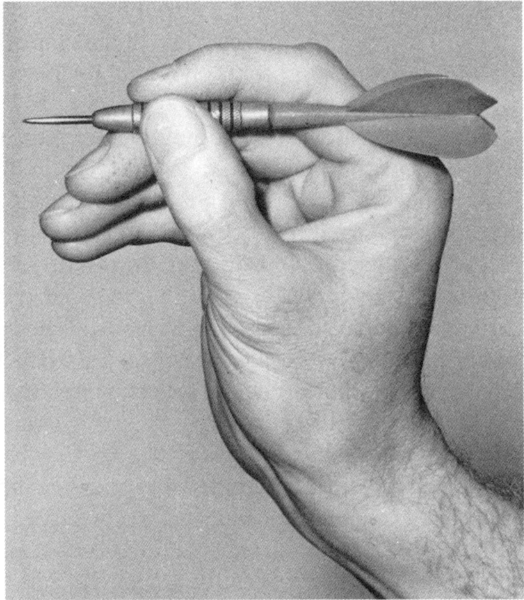

Fig. 25. Opposition in practice. For increased stability and sensitivity the middle finger is brought into action.

tion and feeling; the greater the surface area of highly sensitive papillary skin available, the more effective is the handling of small and delicate objects.

Opposition in Monkeys and Apes

A major factor in achieving pulp-to-pulp contact is the proportionate length of the index finger and thumb. The formula *total length of thumb × 100/total length of index finger* provides a method of expressing the length of the thumb quantitatively (opposability index). A low index denotes a long first finger and a short thumb; and at the other extreme a high index has a rather narrow range in primates, with mean values between 40 and 65. The shortest thumbs and the longest forefingers are found in the great apes, particularly the orangutan (mean: 40). Contrariwise, the longest thumbs and shortest forefingers are exhibited by people (mean: 60). Not all primates are capable of opposing their thumbs. The operative word, you will remember, is *rotation*; without rotation, true opposition is not possible. Only the Old World monkeys and apes show such rotation; New World monkeys and prosimians lack this movement. Thus, the opposability index is applicable only to the Old World primates. New World monkeys and prosimians exhibit what has been called *pseudo-opposability*, a movement in which rotation is entirely absent and pulp-to-pulp contact of the thumb and fingertips is wanting.

It is particularly interesting to note that it isn't the apes—as might be expected—whose opposability index lies closest to ours. The ape hand is almost wholly suborned to a specialized locomotor style of brachiation.* Apes are brachiators, using their long arms and long hands to swing from branch to branch and tree to tree. Brachiation preempted the function of the hand and relegated its manipulative talents to a secondary (although still important) role (Fig. 26). Gorillas, as has been mentioned, are ground-living animals, seldom climbing trees and never brachiating. Although they have not lost the anatomical hallmarks that identify them as *ci-devant* arm-swingers, they have nevertheless become secondarily adapted to a ground-living way of life. The opposability index of gorillas is the highest of all apes, and their manipulative skill with small food objects is by far the best developed. Paradoxically, the gorilla, unlike the chimpanzee, is neither a tool-user nor a tool-maker in nature. According to observers in the field, the gorilla totally ignores alien objects planted to test curiosity and implementation skills.

Much closer to people in manipulative skills, as is suggested by their high opposability indexes (mean: 57–58), are the baboons and mandrills. Although the hand of a baboon is, in effect, a foot in locomotor terms, it has at least several important functions for which finely adjusted thumb-index opposition is essential. These include feeding on seeds and grain and delicate

* Versatile climbing and feeding in the small-branch regions of the canopy are now considered to be important behavioral correlates of morphological features that were ascribed to brachiation and brachiators (see Tuttle 1975, 1986).

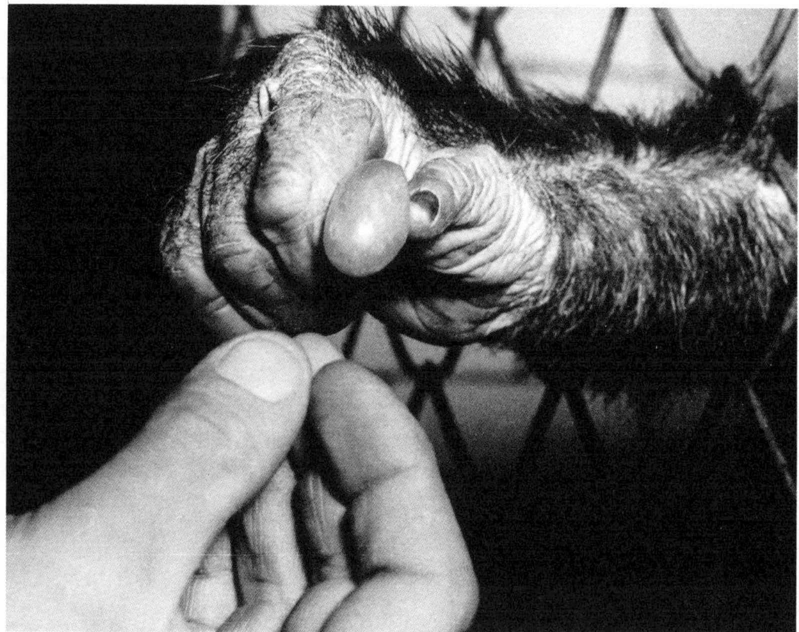

Fig. 26. Chimpanzee grasping a grape in thumb-index opposition. Note the imperfection of the action due to the presence of a short thumb and a very long index finger. Note, too, the "perfection" of human opposition.

shoots of young grass that are plucked with finger and thumb and grooming. Grooming, or fur-cleaning, is both a physiological need and a psycho-social necessity. During grooming, which is performed with great concentration and delicate thumb and index finger work, not only is the skin cleansed of debris and parasites but also the social bonds between groomer and groomee are immeasurably strengthened by a service given and received. It is a matter of "I will scratch your back, if you will scratch mine." A relationship based on such trust and cooperation must make for good rapport between members of a troop to the ultimate benefit of the whole species. Tactile communication between male and female through grooming is also an important element in precopulatory behavior.

Grooming is not restricted to the nonhuman primates; people are inveterate groomers. It manifests itself principally in verbal communication. One can observe "verbal grooming" in action in any pub, hairdressing establishment, coffee-shop, or cocktail party, when, for instance, conventional inquiries by the groomer lead the proud parent (the groomee) to discourse at length on the virtues of his/her offspring. Actual physical grooming also has a reassuring and tranquilizing role to play in people's lives, though such activities lack the concentrated intensity with which they are pursued by monkeys and apes. There can't be much intensity in the classic grooming situation of the wife seeing her husband off to work, straightening his tie and brushing the dandruff off his shoulders.

Movements of the Hands as a Whole

The concept of the *position of rest*, as proposed by John Hilton (1804–1878) and independently by a Gloucester physician named Thomas Ellis, is no longer talked about nor does it figure in the medical textbooks, yet the principles that Hilton and Ellis propounded are put into practice by surgeons all over the world today. The term was introduced by Ellis, but the foundations of the concept were laid by Hilton of Guy's Hospital, whose monograph on *Rest and Pain* (1877) is a classic of medical literature.

Ellis, in his book *The Human Foot* (1889), describes the position of rest as the posture "sooner or later assumed by the limb when in a state of rest from fatigue" or, in other words, during full relaxation or sleep. It is at this time that the outflow of impulses from the brain and spinal cord is much reduced and the muscles are to all intents and purposes fully relaxed. Theoretically the position assumed by the joints is the result of the normal tension of agonist and antagonist groups of muscles in repose, and is such that no part of the ligaments surrounding the joint is on the stretch. The position of rest is the ideal posture (with certain reservations) in which to immobilize an injured joint; it is also the best position should complete stiffness occur, and according to Ellis, in fixing the limb thus we are "doing something toward making complete stiffness the best result to be hoped for." This somewhat equivocal statement is a reflection of the limited opportunities that were available for orthopedic surgery in Victorian times.

Ellis used to watch the limbs during sleep in order to study the position they adopted and then to follow these "lessons of nature." As a result of his experiments, Ellis concluded that the true position of rest "lies somewhere about a mean between the ranges of easy motion" (Fig. 27). Thus the thumb is in the position of rest at the start of the description of its movement (see above); it lies midway .between abduction and adduction, flexion and extension, medial and lateral rotation. The position of the hand at rest has a particular bearing on the rest of this

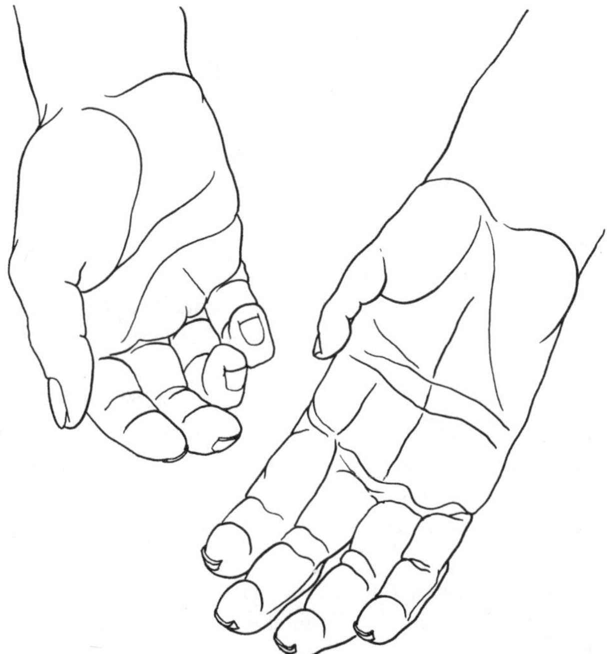

Fig. 27. Human (left) and chimpanzee (right) hands in a position of rest.

section inasmuch as it lies exactly midway between the activity frameworks of *precision grip* and *power grip*.

There are two classes of movement of which the hand is capable: prehensile and nonprehensile. Prehensile movements are those in which an object, fixed or free, is held by a gripping or pinching action between the digits and the palm. Nonprehensile movements of the whole hand include pushing, lifting, tapping, and punching movements of the fingers, such as typewriting or working the stops of a musical instrument. Considering the enormous variety of activities that the hand is called upon to perform, it might be supposed that prehensile movements would be too numerous for simple analysis. However, the diversity of movements is more apparent than real; it is not so much that there is a profusion of *actions* concerned in day-to-day activities as that there is a multiplicity of objects involved—switches, doorknobs, latches, cutlery, cups, glasses, pens, pencils, erasers, buttons, and coins. In fact, there are only two main patterns and two subsidiary prehensile patterns. The main ones are the *precision grip* (Fig. 28b) and the *power grip* (Fig. 28a). The subsidiary patterns are the *hook grip* and the *scissor grip*.

PRECISION GRIP is executed between the terminal digital pad of the opposed thumb and the pads of the fingertips. Large objects held in this way involve all the digits, but smaller objects require only the thumb, the index, and the middle digits. The precision grip is employed when delicacy of handling and accuracy of instrumentation are essential and power is a secondary consideration.

POWER GRIP is executed between the surface of the fingers and the palm with the thumb acting as a buttressing and reinforcing agent. Under some conditions the thumb supplies directional control. Thus, there is a certain element of precision even in the power grip, just as there is an element of power in the precision grip. With heavy tools such as coal-hammers that need little precision in their use, the thumb reverts to its crude reinforcing function, wrapping over the backs of the digits forming a fist. In making a fist, power is the only consideration and accuracy does not come into it.

HOOK GRIP is a subsidiary grip that is a function of the flexors of the fingers; the knuckle-joints are straight and the two terminal joints acutely bent. The hook grip is used when carrying a heavy suitcase,

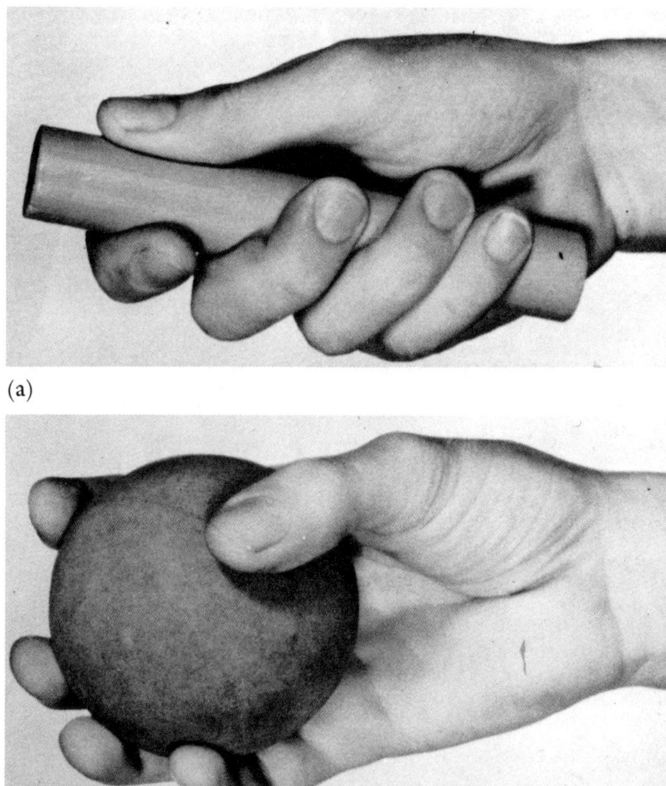

(a)

(b)

Fig. 28. (a) Power and (b) precision grips.

when opening a sash window from the top, or when grasping specialized tools such as pliers or wire-cutters. Although the hook grip is a prehensile posture, it is not a very important one in the general functioning of the hand.

SCISSOR GRIP is a subsidiary grip pattern in which an object is grasped between the sides of the terminal phalanges of the adjacent index and middle fingers. People may grip cigarettes in this

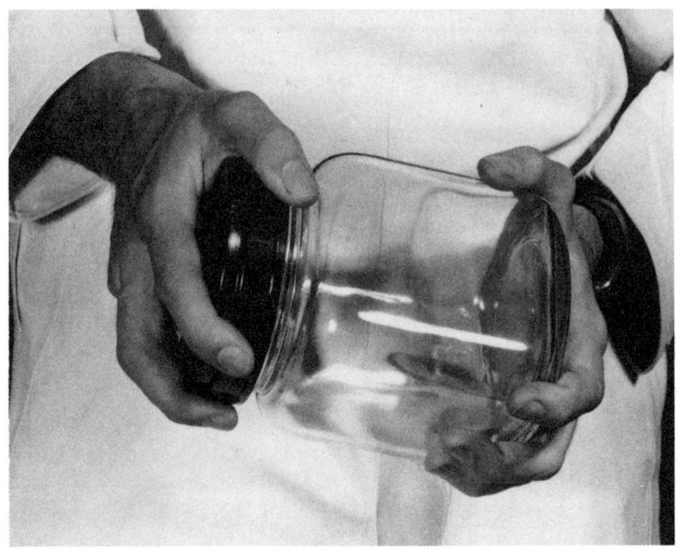

(b)

Fig. 29. (a) Lid of screw-top jar is being "started" with a power grip. (b) Once it is loose, the grip changes to one of precision.

fashion and occasionally they use it casually to pick up small, flat objects when they cannot bring the thumb into action.

It is important to appreciate that the type of grip used in any given activity is *a function of the activity itself and does not depend on the shape or size of the object gripped*. For example, an electric light bulb is held in a precision grip when its lugs are being engaged in the socket (Fig. 30) and in a power grip when the bulb is being screwed or locked home (Fig. 30b). A screw-top jar demands a similar sequence of grips (Fig. 29a and b). The lid is held in a precision grip when the threads are being engaged, but the grip changes to one of power when the lid needs to be screwed home tightly. In these examples the object being manipulated (the light bulb or the lid) remains the same

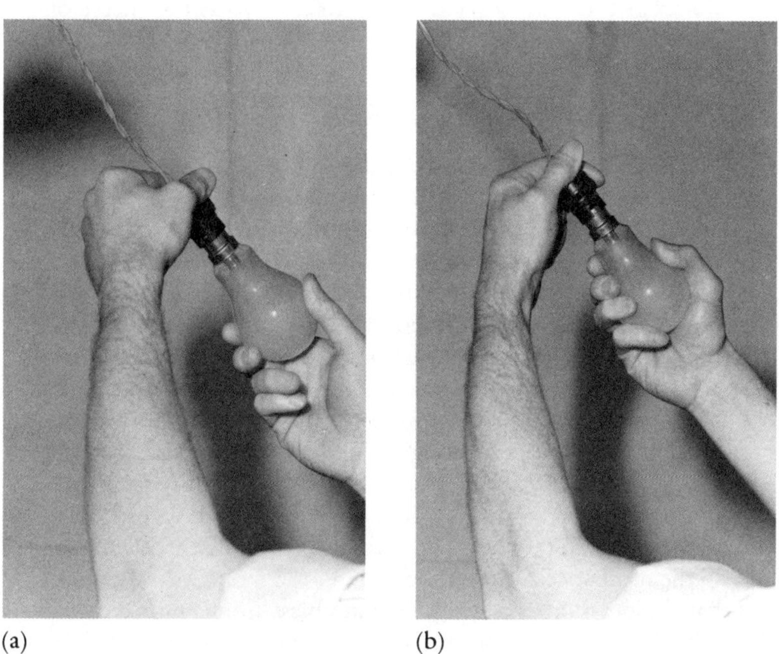

(a) (b)

Fig. 30. (a) Light bulb being engaged into socket using precision grip. (b) Posture of hand changes to one of power grip in order to lock it home.

throughout the sequence; it is the nature of the activity that is changing.

If the object gripped is very small or very large, the generalization may not hold, however. A needle is too small for a power grip and a coal-hammer too large and heavy for a precision grip. In these circumstances there is no choice. The grip may also be affected by the shape of a custom-built tool or handle; for example, a pistol-grip handle with indentations for the digits cannot easily be used in a precision grip.

Precision and power grips are functional concepts, but they are to some extent discrete as far as their nerve supply is concerned. The brunt of a paralysis affecting the median nerve falls on the muscles responsible for the precision grip, so this is the "nerve of precision." The ulnar nerve supplies the bulk of the power grip to muscles and can be referred to as the "nerve of power."

Prehensile Patterns in Primates

Precision and power grips in their fullest expression are hallmarks of humankind, but the evolutionary trend toward the establishment of a dual prehensile pattern takes place at a relatively primitive stage of primate evolution as represented by the New World monkeys. Prosimians (lemurs and lorises) possess only a single prehensile pattern; the hand opens and closes like the jaws of a toy crane in an amusement arcade (Fig. 31). There are no really precise movements and no power movements as such—merely all-purpose grabs. New World monkeys are not much better equipped, though workers who have made a particular study of South American species claim the two grips are distinguishable, particularly in the more advanced species, such as the capuchin monkey. The precision grip is absent, of course, in the most specialized of the New World monkeys—the acrobatic spider monkeys, which are thumbless. Spider monkeys can achieve a modicum of precision by using the scissor grip, in which the object is grasped between the sides of the second and third digits. The prehensile tail of these monkeys is both a power and precision organ. The tail can be used to suspend the body

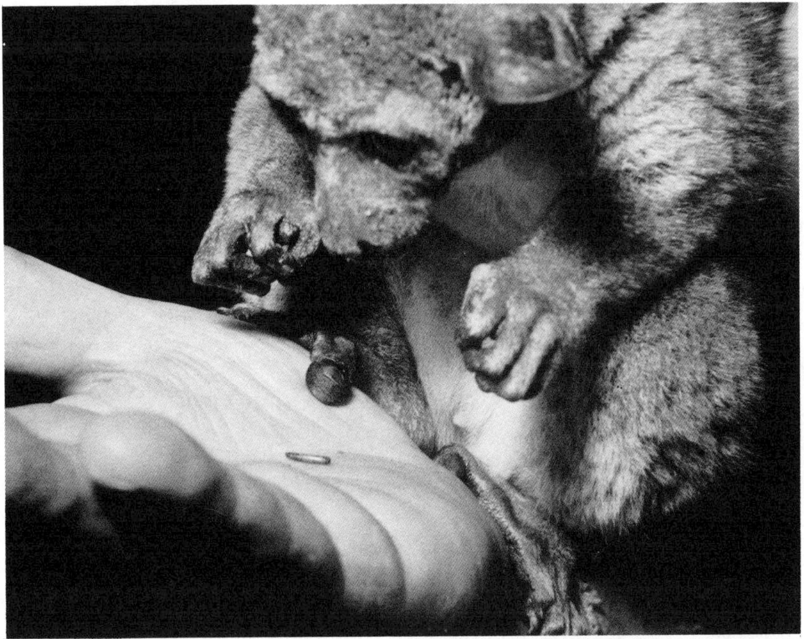

Fig. 31. A galago about to pounce on a mealworm. Note that all digits are prepared for action.

from an overhead branch (power grip), or to delicately pick up such small items as a shelled peanut (precision grip). The undersurface of the distal one-third of the tail is covered by naked skin, identical in function to the papillary skin of the finger.

Old World monkeys are fully differentiated as far as their prehensile grips are concerned, the only exceptions being the colobus monkeys that, like the spider monkeys, have no thumb. Unlike the latter, though, the colobus's tail has no compensatory prehensile power.

The gorilla, orangutan, and chimpanzee show differentiation of the two grips quite clearly, though the precision grip—for mechanical reasons—is less adept in these apes than in some monkeys. Historically, the apes are brachiators and toward this end the hand has undergone a reduction in thumb length and an

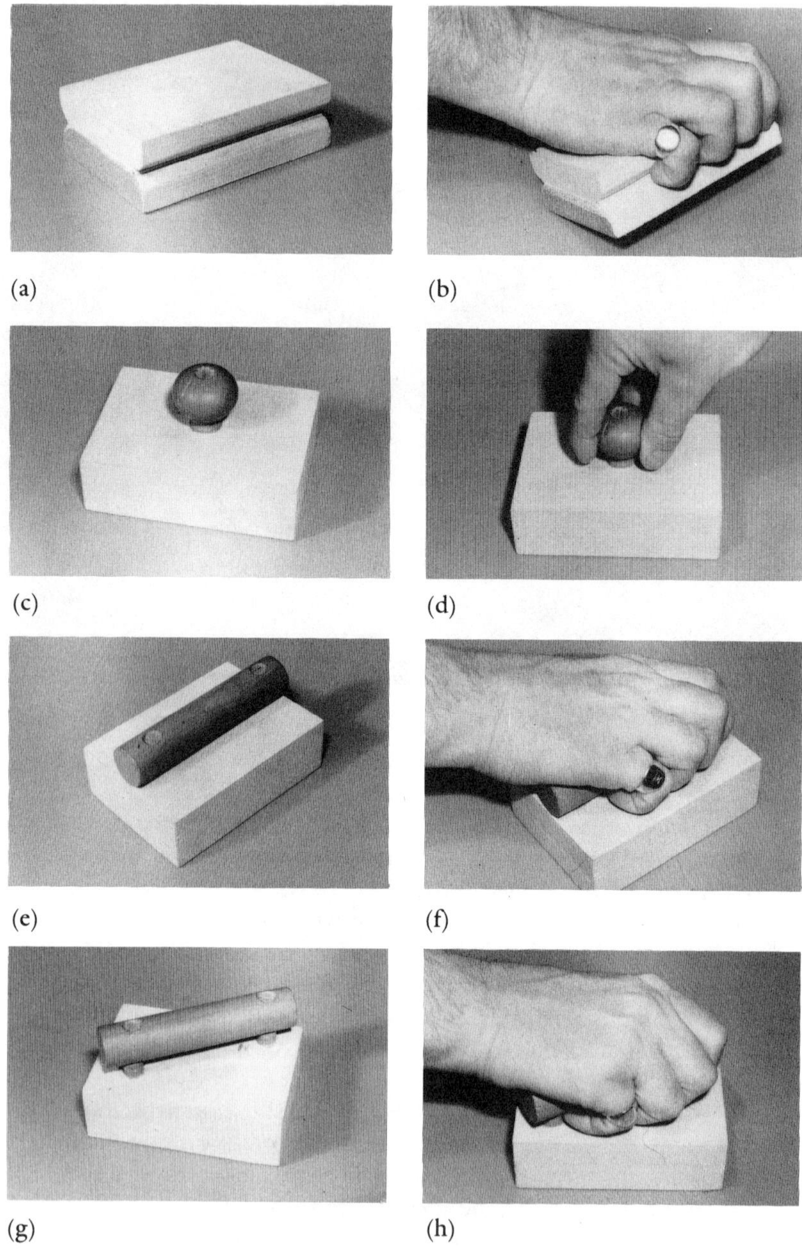

(a) (b) (c) (d) (e) (f) (g) (h)

increase in finger length. Thus finger-thumb opposition (the basis of the precision grip) is considerably impaired (see opposability index, p. 57); occasionally the scissor grip is used.

During brachiation, apes make good use of their elongated fingers, which act as a suspensory hook. The power grip in chimpanzees differs somewhat from ours; the object being gripped is orientated transversely instead of obliquely in keeping with the relatively uncomplicated hinge pattern of the metacarpophalangeal joints (see Fig. 26). This grip is also reflected in the transverse orientation of the simian crease (see Fig. 27). As we have noted, to retain a hook grip is supremely important to apes but is only marginally so to people.

Prehensile Patterns and Tool Handle Design

An extension of the two-grip concept is its application to the design of tool-handles. Most tools and appliances have two ends—the working end and the handle. The form of both components should be dictated by the nature of the particular operation involved, but too frequently it seems that inventiveness and ingenuity are concentrated on the working-end, and the handle-end is left to take care of itself. The human hand is highly adaptable and can secure a grip on almost any kind of shape, but if the tool is to possess maximum efficiency, the handle must be designed for a specific function.

It should be a relatively simple matter to design a handle for a

---

Fig. 32. Theoretical stages in the design and development of a sander, for which a power grip is required:

(a) The first model
(b) Rejected because of inadequate space for fingers
(c) Rounded knob substituted
(d) Rejected since knob provides inadequate surface for power-gripping
(e) Handle substituted
(f) An improvement, but fingers cannot be wrapped around cylinder sufficiently intimately to exert full power.
(g) Handle raised and set obliquely in power-gripping position
(h) Final product satisfactory

single-purpose tool. A careful analysis of the intended activity under all sorts of environmental conditions and situations (cold and hot weather, greasy hands, overhead working, etc.) will determine the most effective grip, whether precision or power. Errors of judgment, however, are still only too common. Let us take the types of nailbrush as examples. One model on the market recently was ball-shaped and made of a shiny plastic material. Not only was the basic shape unsuitable for a power grip (which is the most effective grip to be used when scrubbing the nails) but the rounded, smooth surface, wet with soapy water, was impossible to grip at all. A common contemporary design is the rectangular brush surmounted by a half-hoop in which the fingers are supposed to be inserted; seldom is the hoop wide enough to admit the fingers, and never does it permit the normal staggered flexion of the fingers of a power grip.

Many hand-drills on the domestic market are equipped with a small knob on the driving wheel that is too small to be held in a power grip; a precision grip (quite unsuitable in the circumstances) is all that can be achieved. By means of a simple experiment, I have been able to show that increase in the diameter of the knob until it reaches the dimensions that can comfortably be held in a power grip substantially improves the efficiency of hand-drilling.

The handles of drills and similar appliances are seldom "set" in the ideal functional position for the hand and forearm; as a result such postural corrections as are necessary can only be made by adjustments of the muscles of the shoulder and trunk, which are physically tiring. Fig. 32 shows the hypothetical stages in the design of a tool for the domestic market. There is nothing to go wrong at its working end—sandpaper is sandpaper—but there are plenty of errors that can creep in in the designing of its handle.

Humans have passed from tool-users to tool-makers (see p. 87) and now—somewhat ironically—back to tool-users again. There are a few craftspeople left who make their own tools, but not many; tools are no longer personal creations of craftspeople (whose survival once depended on their effectiveness) but are the standardized products of commercial tool-makers. Tools made for the uncritical domestic market are the worst offenders; shapes of handles are more often chosen for their packaging and their modern design qualities than for their functional suitability. Early people in the throes of constructing hand-axes never made that mistake; their lives and livelihood depended on it.

CHAPTER FOUR

Evolution of the Hand

HUMANS and other primates live what might be called a "hand-to-mouth" existence. In contrast to most nonprimate mammals who put their mouths to the food, primates carry food to their mouths. As might be expected, such a fundamental difference in behavior is accompanied by some notable anatomical modifications, the most outstanding of which are prehensility of the hand, shortening of the snout, and migration of the eyes to the front of the face.

The evolutionary predecessors of primates, the insectivores, constituting shrews and their kind, were long-snouted, nocturnal creatures with high-cusped puncturing and crushing teeth. The sense of smell dominated their way of life; insectivores navigate with their noses. Noses have a tactile as well as olfactory role to play in searching for food. The snout is used to locate and identify the prey, the jaws to seize it, and the teeth to pierce and kill it. Insectivores, with their long snouts and highly developed sense of smell, might be described as "noses on legs." Migration of the eyes to the front of the face, along with refinements in the eye itself and in the visual pathway to the brain, shifted the sensory emphasis toward vision and away from smell. The earliest true primates were vision-orientated and probably diurnal in habit. All this was happening some 65 million years ago.

Primates share their hand-to-mouth feeding habits with a number of mammals that are exceptions to the general rule. Squirrels and other rodents, bears, raccoons, and otters are also hand-feeders, but there are behavioral differences that set the primates apart. For example, marmots and squirrels feed by holding food objects in *both* hands rather than in one (see Fig. 17). The combination of two nonprehensile hands substitutes for the single-handed prehensility of primates.

The matter of the giant panda's notorious "thumb" has already been mentioned (p. 15). The discovery that the panda feeds itself one-handedly would seem to contradict the principle that nonprimate mammals are two-handed feeders, but once again anatomy comes to the rescue of behavior; an enlarged wrist bone mimicking the thumb is a unique adaptation, a one-off event, that does not really affect the general principle.

Arthur, the famous cat-food cat of a British television commercial, appears to be another exception to the two-handed rule, since day after day he can be seen scooping fish out of a can with one paw. However, fish is soft and Arthur's claws are sharp. I submit that Arthur is really a red herring.

Eating has been a prolific fount of social rules throughout the ages, the eighteenth and nineteenth centuries in particular having much to answer for. Which part of the spoon should be used for supping soup? Should bread be cut or broken at the evening meal? Is it incorrect to hold your knife as if it were a pen? Which way should the port circulate?

Table manners and conventions, in maintaining the unity and cohesiveness of the group, serve a very useful social purpose, analagous to the grooming behavior of baboons and macaque monkeys. Manners represent a symbolic reiteration of the current rules of social class and, at the same time, they represent a cultural discipline that insists that the conventions are adhered to by gentlemen even if, like Charlie Chaplin, they are snowbound in the Yukon with only the toughest of leather soles to stave off the pangs of hunger. Carruthers, too, always dressed for dinner in the most putrid reaches of darkest Africa, in spite of incessant drums and restless natives.

However, table manners are not innate, but instead learned, and can be temporarily suspended by group decision as was the case, in a genteel sort of way, with Marie Antoinette and her court. Less histrionic than life at Le Petit Trianon is the picnic where even the most fastidious will gnaw on the leg of a chicken, Charles Laughton fashion, and fling the bones over the shoulder into the nearest bush. To eat alfresco provides the welcome excuse for temporarily rejecting the social rule that food should be eaten with a knife and fork. Humans are hand-to-mouth feeders just like the rest of the primates, but in Western civilizations convention insists that we try to conceal the fact.

Classification of Primate Hands

A monkey's paw has a dual role; it is required to act as both hand and foot. In monkeys a compromise has been reached by which both functions can be served. Human hands have only a single role—that of prehension—and so no adaptive compromise is required and natural selection has had the easier task.

The Foot-Hand

People alone among the primates have fully emancipated their hands from any locomotor role once their childhoods have passed. When young, we crawled quadrupedally with the flat of our hands applied to the ground like a monkey. We did not, as might be expected from our immediate ancestry, knuckle-walk (Fig. 7) or fist-walk (Fig. 33) as great apes do. Instead, as they grow up, human children briefly pass through the foot-hand phase that a monkey enjoys throughout life.

The foot-hand of a monkey, as its name suggests, is capable of both functions. The presence of prominent palmar pads, the cushion of palmar skin on the ulnar side and its extension on to the wrist, forming a sort of "heel" to the hand, provides evidence

Fig. 33. Orangutan and human infants. Note that the former fist-walks while the human crawls in an open-handed fashion, like any monkey.

of locomotor use, while the flattened nails, expanded digital pads, and the mobile thumb bear witness of manipulatory capability. Both sets of characters are equally well developed, with the result that Old World monkeys combine expert handling of objects with efficient locomotion. Naturally, there are limitations to the possession of a foot-hand, otherwise there could certainly have been no selection in humans for bipedalism. The monkey's foot-hand is prevented by locomotor commitments from carrying objects from place to place, while bipedalism leaves the hands free for this purpose. The carriage of food was probably the most significant benefit accruing from the bipedal gait, but the carriage of other objects, such as weapons, tools, and infants, would also have provided undoubted evolutionary advantages for the emerging humans.

The True Hand

Primates with true hands are the apes and humans. Although superficially different, ape and human hands are cast from the same mold. In contrast to the foot-hand of monkeys, the "heel" of the *true hand* is wanting, a clear indication of the absence of weight-bearing on the flat of the hand. The palmar pads have also disappeared as circumscribed prominent hillocks, though the remnants still persist in true hands as reminders of the past. Dissimilarities between ape and human hands are largely quantitative; the ape hand is elongated in its metacarpal and phalangeal segments and the thumb is very short; this is particularly true of the orangutan. The gorilla has the longest thumb and the shortest fingers of all the great apes, and is thereby closest to us in its manual proportions.

While apes never use the hand as a foot on the ground (although the reverse is true), it plays a major part in tree-climbing and suspensory behavior. Nowadays, the African apes are rather infrequent brachiators. Whatever they did in ancient times, they are now found more often on the ground than in the trees; in fact the male mountain gorilla seldom leaves the ground because few trees in his habitat will support his bulk. Locomotion for the gorilla and the chimpanzee is a bit of a problem. Anatomically they are not fully adapted for quadrupedalism nor are they equipped to walk bipedally for more than a few steps at a time. The solution is a sort of compromise, an expedient called *knuckle-walking* in which the weight of the forepart of the body is borne on the middle phalanges (see Fig. 7); thus, since the forelimbs are longer than the hindlimbs, the body is supported in a semi-upright posture. In spite of bearing weight, the hand is not acting as a foot but as a hand; gorillas and chimpanzees, in fact, walk on their hands. Knuckle-walking is a well-established gait associated with a number of special characters of the hand and forearm that make it possible. On occasion humans use their hands to suspend themselves in much the same way as brachiators do, although our size and our weight-to-muscle ratio militate against it, except in the young (shades of those horizontal ladders in the school gym!) or in specially trained acrobats. Further, in special circumstances, humans use a knuckled posture, not to walk, but instead to support the body upright, for example, when standing at a table, delivering a speech.

To sum up, the great majority of primates possess a *foot-hand* that is fitted for both manipulation and locomotion. A minority possess a true hand, which has a dual purpose in apes, but a single purpose only, manipulation, in humans.

Origins of Prehensility

For nearly one hundred years the question of whether mammals were derived from an arboreal or terrestrial ancestral stock has been a key one among biologists. While it is unarguable that all modern primates were derived from an arboreal stock, there is no convincing evidence that the ancestral mammals were.

All primates possess divergent and opposable (in its widest sense) big toes and the question is this: are prehensile big toes an ancient mammalian heritage dating back to the Jurassic and Lower Cretaceous, when the first true mammals evolved, or are they a relatively recent adaptation to life in the trees? If the fossil feet of ancestral placentals were known well enough, the puzzle might be solved, but the truth is that there is no adequate evidence to show if placentals possessed divergent and grasping big toes before the Eocene (see Fig. 34).

The paleontological basis for arguing an arboreal ancestry is

GEOLOGICAL TIME SCALE
TERTIARY AND QUATERNARY

DURATION OF EPOCH (MILLIONS OF YEARS)		
	11	PALAEOCENE
	19	EOCENE
	12	OLIGOCENE
	19	MIOCENE
	3	PLIOCENE
	2	PLEISTOCENE

66 million years—approximate total duration

Fig. 34. Geological epochs during the Tertiary Period.

that *Didelphis*, an opossum-like form, possessed a divergent big toe, but *Didelphis* is a marsupial. Since there is no suggestion that the marsupial grade was a staging-post for all primitive mammals on their way to becoming placentals, the existence of prehensility of the foot in early marsupials is really quite irrelevant. Arboreal the earliest marsupials may have been, but the placental stock, as far as we know, was ground-dwelling—living a nocturnal insect-eating life on and under the forest floor among the roots of the trees. An arboreal background for placentals remains only a theoretical possibility and we must look elsewhere for the origins of prehensility.

A divergent big toe or thumb is usually associated with the power of prehensility—the ability to seize or grasp an object and hold it securely in one hand; but there is more than one way of

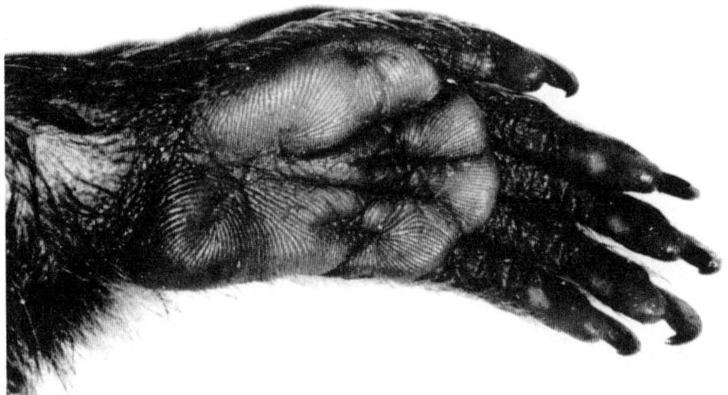

Fig. 35. A marmoset hand. Note palmar pads, narrow finger tips and claws, and little differentiation of thumb. This is a primitive form of foot-hand.

achieving this. For example, neither the spider monkey nor the colobus monkey have thumbs, yet both are capable of effective grasping. The basis of their prehensility is finger flexion. Having exceptionally long fingers relative to the rest of the hand and by folding these digits into the palm, a powerful grasp can be achieved. Orangutans, with a technique of grasping slender supports by means of the double-locking mechanism (see Fig. 11), have remarkably short thumbs and excessively long palms, a situation that approaches colobus hands in its functional proclivities. Even clawed mammals with long digits can achieve a measure of prehensility. Marmosets, which have short clawed digits and a long narrow palm, provide good examples (Fig. 35).

Increase in size and a shift in diet appear to be two of the major factors involved in the evolution of prehensility in the primates. The effect of a change from insect-eating to fruit- and leaf-eating is that a different part of the trees is used for feeding.

Broad-leaved trees, particularly those with spreading or umbrella shaped crowns, are characteristically hollow within. As Defoe said in another context: "Crowns are empty things". Fruit and leaves are found in their greatest abundance on the slender, whippy branches at the extreme periphery of the crown.

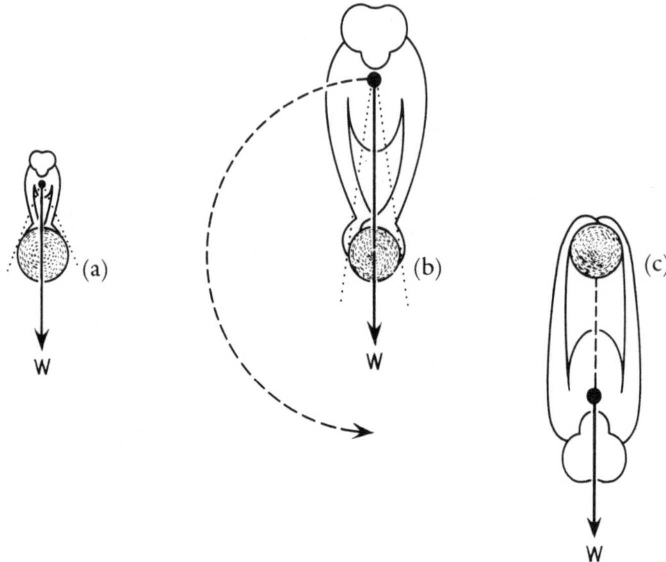

Fig. 36. Relationship between branch size and body size.

This small-branch setting is the feeding zone for the leaf-, fruit- and bud-eaters, and it was this milieu that must have attracted the earliest arboreal primates.

Provided that the ratio of branch size to body size remains high, the axis of gravity is perpendicular to the axis of the branch and the center of gravity is low, arboreal animals can maintain their balance on the branches of trees by means of claws or digital pads alone (Fig. 36a). Locomotion on such branches of trees offers no particular problems since, for small animals, branches provide broad aerial highways theoretically no different from comparable substrates on the ground. With increase in body size a critical point is reached where a potentially unstable system is produced and lateral displacement of the axis of gravity will lead to overbalancing (Fig. 36b). Without some compensatory trick such as prehensility (Fig. 36c), large animals would find it extremely difficult to move about among slender leaf- and fruit-bearing branches and keep their balance; thus, a

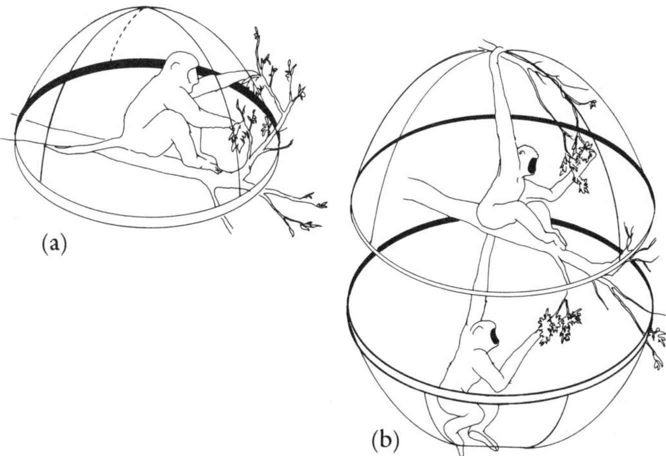

Fig. 37. Feeding kinespheres: (a) Macaque monkey (b) Gibbon. (Courtesy Ted Grand)

rich food source would be closed to them. The rule must have been: stay small or, if you must get bigger, then evolve prehensility at the same time!

A talent to suspend the body while feeding in trees is a characteristic of most primates, though not used by all to the same degree. The adaptive value of suspension is that the feeding range, theoretically, is thereby doubled to include leaves and fruit above and below the branch (Fig. 37b).

The suspended posture may be quadrupedal, bipedal, or bimanual. Bimanual suspension is of particular interest since it is practiced by the gibbons and occasionally by the rest of the apes. It is also a familiar sight to see a gibbon hanging by one hand, the body twisting and pivoting as the long free forelimb reaches out and around to pluck fruit.

The South American spider monkeys, woolly monkeys, and howler monkeys are also habitual suspensory feeders, supplementing their prehensile limbs with a prehensile tail.

The feeding behavior of quadrupedal monkeys, such as the macaques, the langurs, and the guenons, is to sit precariously on slender branches, well anchored by their grasping feet, and to bend back the terminal fruit- and leaf-bearing twigs so that they may be plucked by hand (Fig. 37a). The emphasis in these quadrupeds has not been so much on suspensory-prehension as on supportive-prehension; such a pattern of behavior does not lead to the selection of long arms, hook-like hands, and modifications in the shoulder region that took place among the arm-suspenders. Arm-suspension and arm-swinging have evolved many times in parallel among the primates; particularly striking is the extraordinary morphological similarity between spider monkeys of the New World and gibbons of the Old, both suspensory feeders but otherwise only remotely related. Parallelism, as this evolutionary phenomenon is called, has played a major part in the evolution of the primates.

Thus a dietary change may well have been the initiating factor in the evolution of prehensility, a plausible chain of events that can be summarized as follows. Assuming that the habitat of the earliest primate insect-eaters was the deep litter and low shrubs of the forest floor, the first effect of a change of diet would eventually have been a change of habitat by exploitation of the trees where food was more prolific and of a different sort. Feeding in trees on fruits and leaves would have raised many problems concerned with life high above ground, not the least of which would have been the question of stability. Claws would have been reduced to nails and digital touch pads on the tips of the fingers would have become important. Divergence of the big toe probably preceded divergence of the thumb, but soon competition would have placed strong selection pressures on the increased mobility of the thumb and the manipulative capacity of sensitive touch pads. Thus, the forefoot would have become a foot-hand.

Opposition is a much misused word and has been employed freely to label any movement of the thumb whether or not *rotation* (which is the key movement, see p. 52) is occurring. For example, to describe the movement in New World monkeys as opposition would be quite incorrect, for rotation of the thumb is absent. Since, superficially, the movement of the New World thumb is broadly similar to thumbs that actually do show rotation, the term *pseudo-opposability* is used.

Fossil Hands

The earliest scrap of fossil evidence about primates comes from Purgatory Hill, Montana. It tells us nothing about the hand; in fact it tells us very little about anything beyond the fact that the sometime owner was possibly a primate. The scrap in question is an upper molar tooth, and it has been assigned to a new genus called *Purgatorius*, a genus that was living contemporaneously with the last of the dinosaurs, 60 million years ago. Even to some case-hardened paleontologists, the assignment of *Purgatorius* to the primates on the basis of a single tooth seems presumptuous; it may indeed have been ancestral to monkeys and apes, but there is the distinct possibility that its primate-like shape was merely a functional adaptation to a particular diet and way of life.

The fossil family, the Plesiadapidae, found in North America and Europe, provides much more information about early primate life. They came in a variety of different sizes from rat-size to cat-size; they were ground-living, probably diurnal, and were fruit- and leaf-eaters. Their hands were essentially feet with no specialization of the thumb; on the end of the digits were strong claws. It was possible that some species had adopted arboreal life but had not yet undergone any arboreal adaptations. At this stage—some 60 million years ago—there is no evidence of development of the typical primate foot-hand with its functional traits of prehensility and opposability.

Nonarboreal primates, such as the Plesiadapidae, were quite likely to have come up against the strong selection pressures of rodent competition in the ground-living niche that they both shared. For many years it was a popular, if somewhat simplistic, theory of primate arboreality that the evolving rodents in effect pushed the primates up into the trees by preempting the available niches at ground level. I am satisfied that this is one reason at least why primates became arboreal, though not the sole one. A preference for edible tree products (including insects) was undoubtedly another compelling factor; an insectivorous element figures in all primate diets.

The Plesiadapidae were the last of the early ground-living

primates. From that time on primate evolution took place in the trees, the only environment in which the major distinctions of the primate hand could have developed. Paradoxically, certain refinements of the human hand evolved in quite another milieu, but more of that later.

By the Middle Eocene (see Fig. 34), an important group called the Adapidae were flourishing in North America and Europe. Two American genera, *Smilodectes* and *Notharctus*, are well known from fairly complete fossil skeletons. The adapids were rather like modern-day lemurs, with hindlegs rather longer than forelegs and slim acrobatic bodies. For the first time in the fossil record we find in creatures of an arboreal habitat that not only are the digits surmounted by nails instead of claws, but also the hands are prehensile. The fingers are long and the thumb is mobile, probably pseudo-opposable.

Toward the end of the Eocene, a group of primates emerged whose descendants, the tarsiers, are still with us today in Southeast Asia and the Philippines. The extraordinary thing is that the Eocene tarsier and the modern tarsier are virtually the same animal, made up of the same components. We can safely assume that the hands were identical with their modern counterparts. Instead of two distinct prehensile patterns found in monkeys and apes, the tarsier has only one pattern, which is neither a precision grip nor a power grip; for want of a better word, and at the same time to illustrate the catch-as-catch-can nature of the movement, we can call it a grab. Galagoes show much the same grabbing movement (see Fig. 31), and indeed it is likely that by the end of the Eocene no precision or power grips as defined had yet evolved in the primate stock.

The Oligocene was the geological epoch when the monkeys and apes first appeared ($\geqslant$30 million years ago). The site of the most important fossil remains is a region called the Fayum in Egypt, which was at one time an area of lush tropical forests well supplied with rivers and swamps, now an arid waste of desert sand. Although skulls, jaws, and numerous teeth have been found there, fewer limb bones have turned up, so this epoch is a blank as far as hand evolution is concerned. However, working forward from the Eocene and backward from the Miocene (when hand bones are known) we can build up a picture of a hand with fairly long fingers, a pseudo-opposable thumb, sensitive touch pads on the fingertips, and the beginnings of power and precision grips, much like the hand of New World monkeys today.

The first unequivocal ape ancestor that we know of lived 19 million years ago, was called *Proconsul* and was about the size of a fox-terrier. *Proconsul* fills a very important gap in the fossil record because it represents one of the many kinds of apes that lived before the "great divide" took place—the dichotomy between apes on the one hand and ancestral people on the other. For many years it was thought to be the common ancestor, the progenitor of the ape-human stock. *Proconsul*, discovered in East Africa in 1948 by Drs. L.S.B. and M. D. Leakey, was undoubtedly an ape (not a monkey) whose teeth were very similar to a living chimpanzee's. Surprisingly, below the neck it was more monkey-like than ape-like; unlike apes, *Proconsul* was not a brachiator but a fairly typical quadruped who ran, climbed, and leaped among the trees, but did little arm-swinging. The characteristic gait of apes evolved later and—as already pointed out—led to a rather specialized sort of hand. Protohumans, who were never true brachiators, though they may have done some arm-swinging in their evolutionary youth, at the common ancestor stage, perhaps, avoided the pitfalls of the specialized brachiator's hand.

Fortunately a single specimen of the left hand of *Proconsul* was discovered in 1950 in an infilled pothole in the floor of an ancient buried lake; most of the arm bones were found with it (Fig. 38).*

The hand was flat-nailed, prehensile, and had a pseudo-opposable thumb. The general shape was a cross between a monkey's hand and a chimpanzee's. The metacarpal heads were smooth and globular and totally lacked the sesamoid gutters that are so characteristic of Old World monkey hands; the carpal bones were ape-like in their shape and function, except for the

* More of this individual (KNM-RU 2036) and many additional specimens of *Proconsul* have been recovered from Kenya during the past decade.

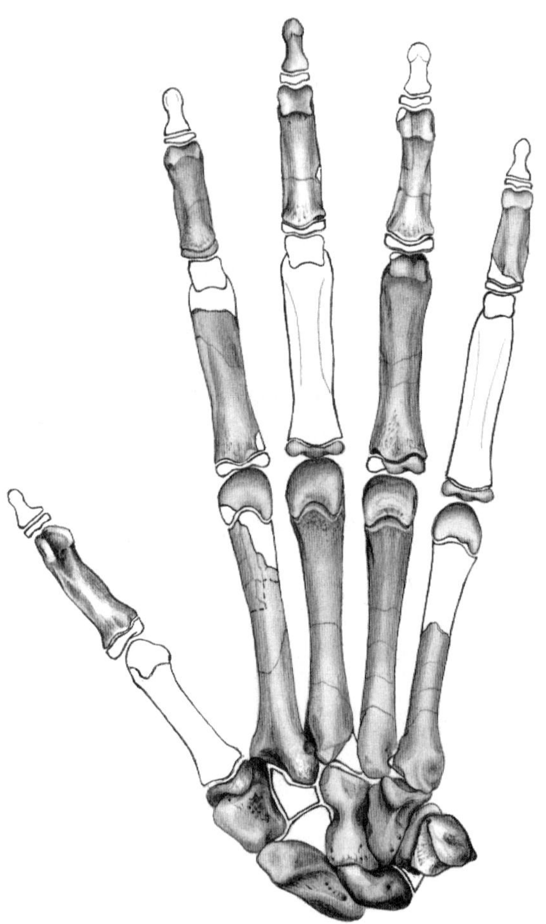

Fig. 38. Partly reconstructed hand of *Proconsul africanus*, a dental ape from Kenya, once thought to be especially related to the chimpanzee, but now believed to be too old for that role (19 million years BP [before present]).The proportions of the hand are not chimpanzee-like; the thumb is not truly opposable.

trapezium—the bone that articulates with the base of the thumb. Instead of a saddle-shaped surface denoting true opposition, the joint was cylinder-shaped (as in modern New World monkeys), indicating pseudo-opposition. *Proconsul*'s hand was monkey-like in overall proportions inasmuch as the fingers were rather short and the thumb rather long, the proportions being the reverse of the chimpanzee's hand.

One must assume that when the descendants of *Proconsul* turned their backs on an arboreal life and started the long evolutionary trek to humanhood through life on the ground, the hand was still a foot-hand. Only with the assumption of an upright posture and a bipedal walking gait would natural selection have favored the hand as a nonweight-bearing multifactorial instrument, gradually eliminating the characteristics of the foot-hand (the palmar pads and "heel" of the hand, for example) and replacing them with expanded fingertips, a stout and truly opposable thumb, and a broad palm.

Unfortunately from the time of *Proconsul* to the era of the australopithecines, the African near-humans—a time interval of perhaps 16 million years—we have no record of the hands.

Nature has been much more giving of hands from Pliocene protopeople (Fig. 40). A veritable treasure trove of hand bones from Hadar, Ethiopia, show that 3 million years ago *Australopithecus afarensis*—the formal name for "Lucy" (A.L.-288) and her kith and kin—had small, but strong hands that were well adapted for power-gripping sticks and stones, perhaps for vigorous pounding and throwing. However, their precision grips may not have been completely humanoid, and no stone tools are associated with them. Indeed, stone artifacts do not appear in the fossil record until 2.5 million years ago, albeit at another site (Kada Goba) in the Hadar region.

The hands of the South African australopithecines are known from fewer bones, one of the most instructive of which is a thumb metacarpal complete and undamaged from Swartkrans in the Transvaal. There was a good saddle joint at the base of the bone and one can take it that the thumb was opposable. The bone itself was exceptionally robust and the strong muscular markings indicated powerful action of the thumb. In contrast, the parts of two other metacarpals that are known are finger metacarpals, slenderly built and devoid of muscular markings. It seems most likely that the thumb metacarpal belonged to one species of australopithecines (*Australopithecus robustus*) and the finger metacarpals to the other (*Australopithecus africanus*).* The former species, a tough-food vegetarian, was cranially robust, while the latter was a more delicately boned, gracile creature who was perhaps a hunter of sorts and certainly a tool-user, if not a tool-maker. Both species were bipedal.

In 1960 at Olduvai Gorge, Tanzania, Louis and Mary Leakey made an important discovery. A number of hand and foot bones, two skull fragments, a juvenile jaw bone, and a collar-bone were found in Bed I (the lowest of the strata exposed on the face of the gorge) associated with a well-defined "living floor" including tools of a simple form. It was this latter material that led to naming the fossil *Homo habilis*, the "handyman".

The Olduvai hand assemblage originally consisted of twenty-one bones, and was first described by me in 1962. Seven of the bones turned out to belong to some nonhuman species, and one was not a hand bone at all. Of the remaining thirteen, two were adult** and eleven were either juvenile or indeterminate (probably juvenile) in age. Most of the juvenile bones belonged to the right side.

The hand is quite robust considering the age of the individual. The bones lack the slenderness of a chimpanzee or an orangutan; nor do they possess the sheer craggyness of the adult gorilla hand. Morphologically, the Olduvai hand cannot be closely matched with any human or ape species living today; its nearest counterpart would be a combination of features of adult human and a young gorilla (Fig. 39).

The most striking human features are the breadth and potential power of the terminal phalanges, particularly of the thumb, which clearly carried a broad flat nail. In combination with a

* On the basis of further excavations at Swartkrans, it is now believed that only *Homo* and *Australopithecus* (= *Paranthropos*) *robustus* were present. The hand bones cannot be associated unarguably with one versus the other, and both may have used tools.

** The two adult bones also are nonhuman (Day, 1976a).

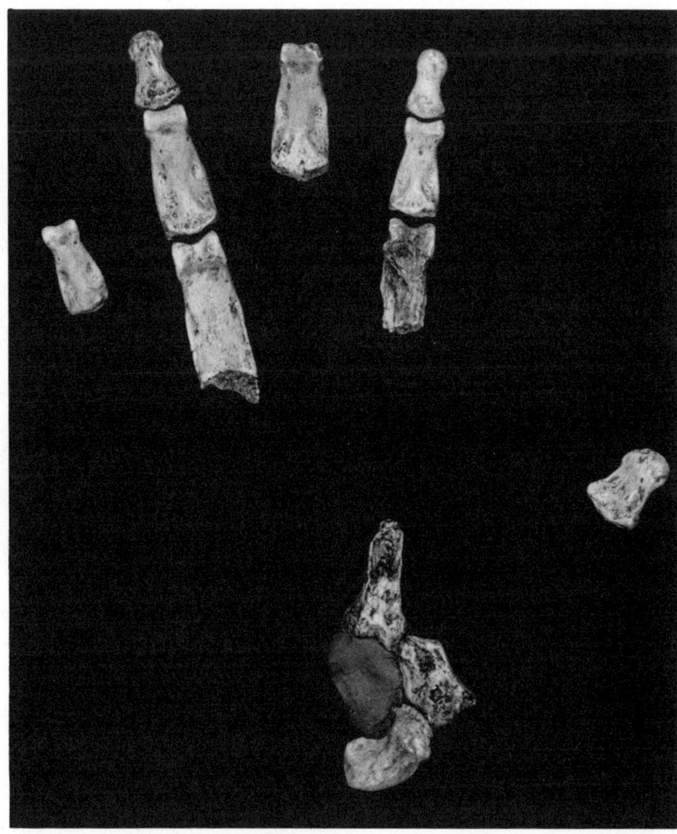

Fig. 39. The hand bones of *Homo habilis* found at Olduvai Gorge, Tanzania, in 1960. Assessed as being 1.75 million years BP, they were found at the same level as simple stone tools called pebble-choppers. Thumb fully opposable. (Courtesy of M. H. Day and J. L. Scheuer)

very well developed saddle joint, the thumb gave the appearance of a notably strong digit.

Functionally, it is very probable that the power grip was well developed and effective but there is some doubt about precision grip that, while undoubtedly possible, may not have been as fully evolved as in present-day humans.

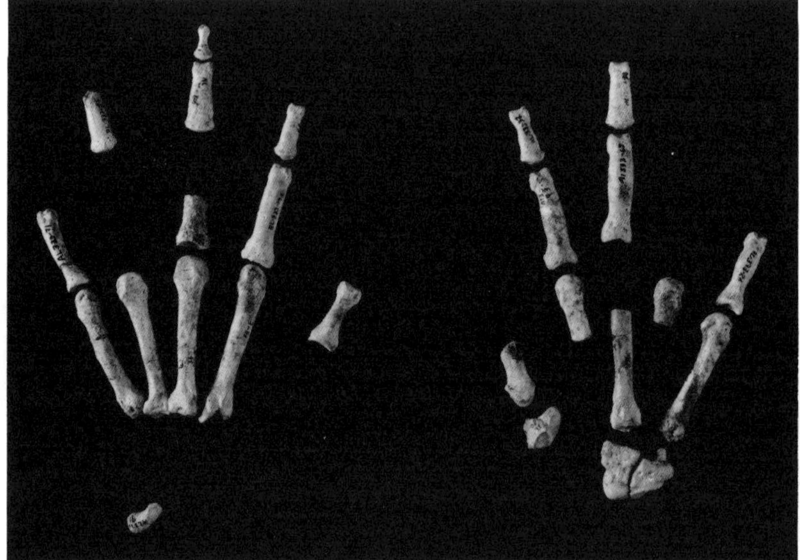

Fig. 40. Fossil hands from Afar, Ethiopia. 3 million years BP. (Courtesy of Michael E. Bush, Cleveland Museum of Natural History)

The discovery of *Homo habilis* at Olduvai Gorge revolutionized our preconceived ideas of the time-scale of human evolution. During the 1950s we scoffed at the idea that humanoids were as much as a million years old, but by 1961 we were in possession of the astonishing fact that, indeed, they were nearly twice as old, 1.75 million years old in fact (*Homo habilis*). Further, in 1973, Richard Leakey produced good evidence that the skull designated KNM-ER 1470 from Lake Turkana, Kenya, is clearly *Homo*, and it bears a potassium-argon date of approximately 1.87 million years BP.

Apart from a single wrist bone of Peking hominids (*Homo erectus*, a half-million years old), no hand bones are known from the time of *Homo habilis* at Olduvai Gorge 1.75 million years ago until the age of Neanderthal people of Europe and western Asia, the ice-beleagured side-branch* of the human race, 35,000 to

130,000 years ago. The villains of the piece largely responsible for this shortfall, and the worst enemies of paleontologists, are the animal scavengers that eat dead flesh, dismantle the limbs, and crunch the bones; acid soils and bacteria take care of the rest. It's a wonder that paleontologists have anything to work on at all.

Mercifully for our knowledge of the postcranial remains of Neanderthalians, the custom of burying the dead evolved during Mousterian times. Cave-living, too, probably offered some protection from scavengers. Whatever is the explanation, Neanderthal hands are quite well known and several informative studies have been made.** One, using the methodology of multivariate analysis, was able to show that the hand bones had certain unique anatomical features. Jonathan Musgrave was also able to show that the degree of dexterity in Neanderthal man was probably not as high as in modern people.

Neanderthal man is everyone's idea of a human ancestor (though he was almost certainly not, being a side branch and not ancestral to modern humans). He is the original caveman of popular imagination, with a squat muscular body and a brutish face. For good measure he carried a knotted club, fought off attacks by dinosaurs and dressed his women in furry bikinis.

In fact, anatomically speaking, he was a good, large-brained sapiens type differing only slightly from modern people. It must not be forgotten that Neanderthalians, during the glacial, were living effectively within the Arctic circle and it is to this ecological fact that they owe many of their physical novelties. Neanderthalians, in fact, were a very spectacular race of "eskimos" with short, sturdy bodies for conserving heat, large knobby joints, and massive jaws that could chew tough, partially cooked food.

The hands themselves (Fig. 41) seem to reflect in every way

* The place of Neanderthals in human evolution is quite controversial (see Lewin, 1989; Mellars, 1989; Klein, 1989; Smith and Spencer, 1984; Trinkaus, 1989).

** Musgrave (1971, 1973, 1977), Trinkaus (1973), Trinkaus and Villemeur (1991)

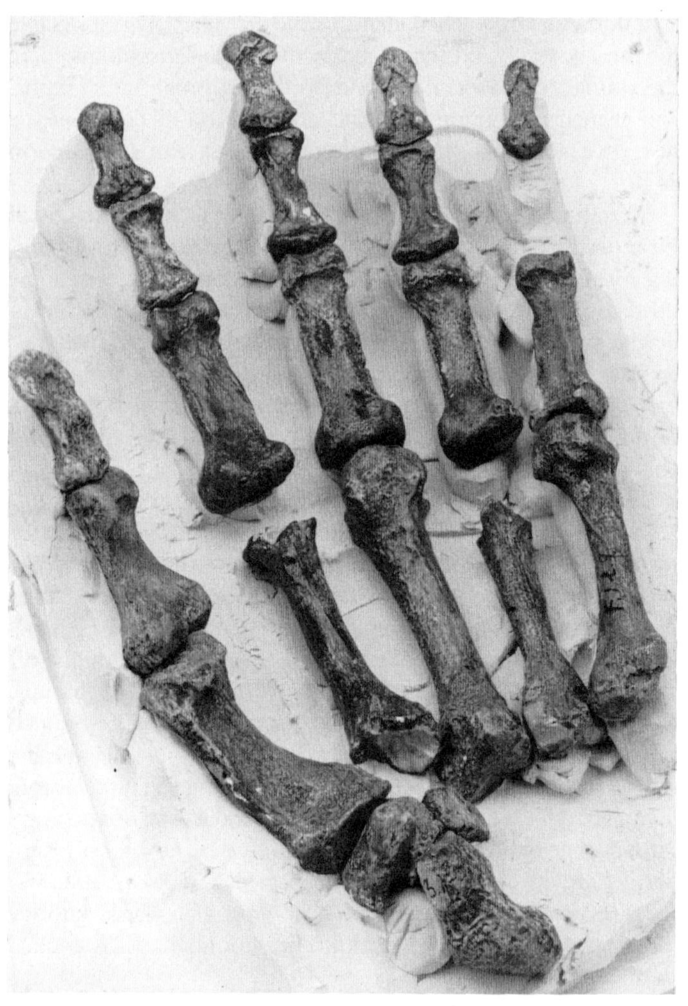

Fig. 41. Fossil hand of a Neanderthal from La Ferrassie, France, 60,000 years BP. Thumb is opposable but the hand is not thought to have been as dexterous as a modern human's.

what we know of Neanderthalian ecology and habits. Like the build of the body as a whole, the hands were broad, the knuckles and joints knobby, the muscles exceedingly well developed, especially the thenar and the interosseus muscles, and the fingertips large and spatulate. In the matter of function there is good evidence to suggest that the precision grip was powerful and well developed. According to Musgrave's estimates based on the left and right hands of the skeleton from La Ferrassie in France, the opposability index (see p. 57) was 64, which lies in the middle of the range of modern humans (61–69). From these measurements there is no doubt that Neanderthalians were able to approximate their thumb and index tips, but just how much and for what they utilized the precision grip during his normal life-style is quite another matter.

Neanderthalians were hunters and their principal chores were to dismember and cut up the carcasses of the animals they killed and to clean and cut up the hides. To this end they developed a highly specialized tool kit of cutters, piercers, and scrapers. In addition Neanderthalians employed more general tools such as hand-axes, spear heads, and single-edged stone knives, but undoubtedly the Neanderthalians' main contribution to the advance of technology was the enormous variety of flake tools, which French archaeologist François Bordes stated to be at least sixty in number. All such tools as these are capable of being manipulated in a power grip—indeed bearing in mind the use to which they were put, they could hardly have been employed any other way.

It was not until the Upper Palaeolithic that true "precision tools" like blades and engravers were used for working bone, antler or wood. This does not necessarily rule out the capability of precision-gripping as such in Neanderthalians; it simply implies that climatically and ecologically hard-pressed as they were, they had no time to spare from the business of survival to indulge in the fripperies of their more leisured successors.

PART TWO

Social and Cultural Aspects of the Hand

CHAPTER FIVE

Tool-Using and Tool-Making

BENJAMIN FRANKLIN was not only a distinguished writer and publisher, a prolific inventor of scientific apparatus, and a wise diplomat, but he was a philosopher and humanist. Perspicacious as he was, it still comes as something of a surprise that it was Benjamin Franklin who was the first to perceive an essential element of human cultural nature: "Man," he said, "is a tool-making animal."

As Kenneth Oakley of the British Museum (Natural History) has observed, the question of drawing a boundary between pre-humans and humans had not yet become a practical issue in Franklin's time. Humans may have existed in the past but that they should have been the subject of evolutionary change was unthinkable. It was not until the last quarter of the nineteenth century, after *On the Origin of Species* (1859) and the *Descent of Man* (1871) had been published, that human thoughts expanded enough to encompass the revolutionary truth that Bishop Ussher might have been wrong about the age of the world, and that, on the contrary, humankind was extremely ancient and must have possessed forerunners who were less human than we are. Even so, the existence of a prehuman or ape-man stage was not so readily accepted until Piltdown man came along to convince a skeptical and anthropologically naive world that the missing link had been found with the unanswerable combination of a human skull and the jaw (suitably tinkered with) of an ape. Tools (several flint flakes that in the event turned out to be faked) did not raise the dust to any extent, so the significance of tool-making was not an issue of great moment.

It was not until the discovery of fossil sites in South Africa that the issue really came to a head. Were the australopithecines that Dart had been collecting so assiduously since 1924, when the first discovery was made at Taung, human or prehuman, or ancient ape?

The first intimation of stone tools in South Africa came in 1953 when pebble tools of an early Oldowan type were found in the two-hundred-foot terrace of the Vaal River. The question was immediately raised as to who the makers were—australopithecines or some more advanced group? In 1948, teeth more humanoid than *Australopithecus*, were found at Swartkrans (where the robust australopithecine was already known) and given the name of *Telanthropus*. Here, then, was an "advanced group" who were tacitly credited with being the tool-makers. *Telanthropus* was subsequently dropped as a generic name when the fossils were identified as human.* Few scientists believed that *Australopithecus* was a tool-maker, principally on the basis of brain size (which was no greater on average than that of the apes); but the power of suggestion had something to do with it. People are the tool-makers; therefore if australopithecines weren't human, they weren't tool-makers. Although remains of australopithecines have been found with Oldowan tools, it is commonly assumed that *Homo habilis* was the tool-maker However, Susman (1988a,b) argues that South African robust australopithecines were tool-makers.

It is perhaps rather surprising that a bipedal hominid with at least as big a brain as an ape's, was not a more advanced tool-user; but so it seems. Australopithecines may have relied solely on improvisation, picking up naturally occurring objects when they found them to fulfil an immediate purpose; or they may have specialized in adapting tools from animal bones for such purposes as digging, pounding, sawing, and chopping. The osteodontokeratic (bone, tooth and horn) culture, as it is called (see below), is no more advanced in the technology of tool use than chimpanzees, which employ sticks as probes and leaves as sponges and "toilet paper."

The distinction between tool-using and tool-making is of critical importance. One might say that almost any primate can use a tool (in the widest sense of the word) but only a human being can make one; the terms must therefore be used circumspectly. The difference between using and making is largely

* Now Telanthropus is arguably *Homo habilis* or *Homo erectus*.

an affair of the central nervous system and involves a qualitative shift in cerebral activity from percept to concept. Abstract thinking is not a talent of nonhuman primates, which live on a strictly "here and now" basis and for whom the past and the future have very little meaning. Some inkling of the power of planning ahead seems to be possessed by chimpanzees, whose tool-modifying activities suggest that they occupy the grey area within the spectrum of tool-using to tool-making—a stage of dawning comprehension.

Imagination is basic to tool-making. All human-made tools start off as chunks of undifferentiated material, which are then shaped according to some cerebral blueprint of what is required. Every stone tool is an act of imagination and—for Aristotle—of artistry, for, as he expressed it in *De partibus animalium*, "Art . . . consists of the conception of the result to be produced before its realization in the material."

The definition of humans-the-tool-makers is a good one, partly because it forms the basis for an objective definition of humans but mainly because it recognizes that behavior as well as structure can be used, even in paleontology, for the identification of *Homo*. Tool-making is a key function in a given fossil in that it supposes the presence of a number of contingent human characteristics such as the power and precision of the hands, bipedalism, and a large brain.

The living great apes, whose hands are classed as true hands, never had a chance to deploy them as people have done. When arboreal, the ape hand was preoccupied with climbing, maintaining balance, and food-collecting; when on the ground the ape hands are still subordinated to the needs of locomotion (knuckle-walking in African apes and fist-walking in orangutans).

The most widely held opinion regarding the splitting of the African apes and human stocks is that this critical event took place before the specialized "brachiating" characteristics of the ape hand had evolved. Russell Tuttle, who holds this view, believes that before becoming bipedal, ancestral hominoids used their hands in a palmigrade fashion as do monkeys, and not in the knuckle-walking posture of modern African apes. He also believes that the early hominid hand had the general proportions of modern humans and not those of the long-handed, long-digited, and short-thumbed apes. Tuttle proposes that selection for a specifically modern type of human hand came with tool-making. I find little in this view to disagree with.

The use of tool-making as a criterion of humanity carries the further implication that its evolution was a sudden event, an inspired flash of conceptual thought, a bolt from the Pliocene blue. It suggests that overnight a divine inspiration lit upon some favored ape-man that thereby set her or his tribe on the way to the stars. But it doesn't work like that. Evolutionary changes are gradual, not abrupt like instant mashed potatoes. Toolmaking grew out of tool-using over millions of years. Study of the use of tools among animals makes it clear that given the physical means and the environmental incentive, many animals, including primates, will use tools. The custom is both widespread and ancient, and it was perfectly logical that the early hominids should have joined the ranks of tool-using animals, given the criteria of physical means and environmental incentive that they undoubtedly possessed. *Humans were tool-users before they were tool-makers*. This is certain and there is no possible way in which such an apprenticeship could have been waived. Furthermore, when they had progressed beyond simple tool-using, they were likely to have passed through a bridging stage, which can best be called *tool-modifying*. Tool-modifying, as its name implies, involves adapting an existing object by simple means to improve its performance. The osteodontokeratic (bone, tooth, and horn) culture of Dart may provide an excellent example of tool-modifying.*

We have now introduced the three grades of a developing technology: tool-using, tool-modifying, and tool-making. Let us define them:

I. TOOL-USING: Tool-using is an act of improvisation in which a naturally occurring object is utilized for an immediate purpose, and discarded.

* Fewer experts accept the osteodontokeratic culture as evidence for australopithecine tool use. Instead, the animal bone deposits could have been accumulated and altered by hyenas (Brain, 1981).

II. TOOL-MODIFYING: Tool-modifying consists of adapting a naturally occurring object by simple means to improve its performance: once used it may be discarded or retained.

III. TOOL-MAKING: Tool-making is an activity by which a naturally occurring object is transformed in a set and regular manner into an appropriate tool for a definite purpose.

Inherent in this definition of tool-making is the maintenance of continuity in successive generations by means of example and demonstration. Whether this is what is meant by culture I am quite unable to say; certainly it is one aspect of it. If culture includes imitation as this brief extract from Goodall (1968) attests—"In the wild I saw infant chimpanzees, on many occasions, not only watching adults as they worked [at termite-fishing] but also picking up and using the same tools when the adults moved away" (p. 209)—then chimpanzees have a culture. Goodall observed the use of leaves as "lavatory paper" by infant chimpanzees in imitation of their mothers' toilet behavior.

Ironically, human tool-making must have been preceded by a lengthy period of ad hoc tool-making and, even earlier, of ad hoc tool-modifying before the technique settled down as part of an established behavioral repertoire, lost and rediscovered many times over, perhaps becoming permanent only with the evolution of speech and language.

Tool-making skill depends on both peripheral and central factors. It depends on the proportions of the hand as well as the size and complexity of the cerebral cortex. S. L. Washburn of Berkeley has emphasized that increase in brain size (a somewhat crude but useful method of estimating the overall capability of the brain in terms of motor and tactile functions, skill, memory, and foresight- all of which take up brain space) is more likely to have followed than preceded tool-making, so that a positive feed-back became established. A breakthrough in tool-making is followed by an increase in the size and complexity of the cerebral hemispheres, which is in turn followed by further advances in tool-making.

Clearly, though, if we want to scramble back nearer the origins of primate and—specifically—human tool-use, we must recall the ancient possessions of, first, stereoscopic vision and prehensility, which provide the basis for hand-eye coordination that is so pivotal to the development of manipulative procedures; and, second, the long record of truncal uprightness which has been a feature of primate locomotion and posture since Eocene times, millions of years before the advent of bipedalism. These possessions provided the groundwork—the preadaptations as they are called—for bipedalism and tool-making. Thus, it was not merely inevitable; tool-making, when it evolved after a long period of incubation, was the culmination of an ancient primate trend involving the hands, the eyes, and the brain in three-way coordination.

Tool Use in Animals*

The classic example of tool-using in birds is the case of the woodpecker-finch of the Galapagos Islands, which makes use of a thorn or a cactus spine held in its beak to winkle out grubs and other insects from the bark of trees. The woodpecker-finch is said to use a certain amount of discrimination in choosing a spine, rejecting those that are too short or too flexible.

Among mammals, the sea-otters of the Californian coastal waters are famous for their appealing method of feeding. Floating on their backs they employ a flat rock balanced on their chests to act as an anvil for cracking open the hard shell of certain mollusks that form part of their diet. Holding a shell-fish in a two-handed grip the sea-otter smashes it with powerful strokes on the rock.

There are many examples of the "anvil" technique in nature. Thrushes, blackbirds, and starlings, for instance, hold an occupied snail shell in the beak and repeatedly beat it on paving stones or an asphalt path until the shell is fragmented and the snail can be extracted. The technique of the lammergeier (the bearded vulture) differs only quantitatively. When it needs to crack open a mammalian long-bone for its contained marrow, it drops the bone in a sort of bombing run on to a large rock from several hundred feet in the sky. An example of tool-using—

* See also Beck, *Animal Tool Behavior* (1980).

which was in no sense an innate action, as were the examples quoted above—was the saga of Jackie. His actions reflected a certain amount of insight in problem-solving. Jackie, an old and toothless capuchin monkey in the London Zoo, was inordinately fond of nuts, particularly brazils. Using a heavy marrow bone in his hands, Jackie, with great persistence, would attack the nut until eventually it succumbed. Marrow bones do not figure much in a wild capuchin's daily round so there was nothing innate about Jackie's action. At his level he was just as brilliant an inventor as Arkwright or Edison. A not dissimilar action has been reported for a capuchin in the wild (Fig. 42). It is important to realize that—as Jane Goodall has said in her book *In the Shadow of Man*, it is only because of the close association in people's minds of tools and humans that special attention has been focused upon any animal able to use an object as a tool. No special intelligence is being displayed by sea-otters, woodpecker finches or lammergeiers. Like Annie Oakley in the song, they are simply doing what comes naturally.

It is among the apes that definite advances in tool use are seen. Yet, oddly enough, it is only chimpanzees that show a real flair. Gibbons and gorillas show no interest in "objects." Dr. George Schaller, the first scientist to study gorillas at close quarters in the wild, reports that wild gorillas totally ignore objects purposefully planted in their path. Young orangutans in captivity, however, show a keen curiosity for mechanical objects and usually end by taking them apart; there is little knowledge of the implemental behavior of orangutans in the wild,* although the use of sticks as missiles, levers to open bees' nests, and weapons to kill snakes has been reported. Chimpanzees are a class apart as far as tool use is concerned. Although anatomically very close, chimpanzees and gorillas are essentially and temperamentally disparate, chimpanzees being noisier, more extroverted, and infinitely more inquisitive. These aspects of their temperaments are reflected in the organization of their respective social systems. Stretching a point a bit, one might liken gorillas to a well-brought-up Victorian family dominated by Papa. Chimpanzees

* See Galdikas (1982).

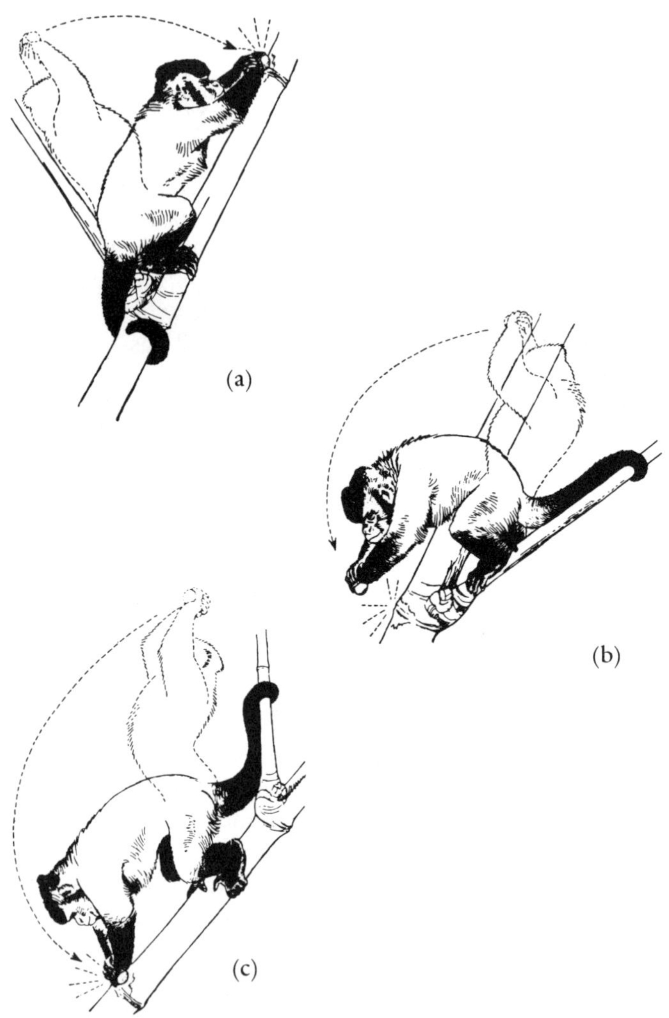

Fig. 42. Capuchin using three different methods to crack open strong fibrous husks of *Cumare* fruit. (Courtesy of Drs. Kosei Izawa and Akinori Mizumo)

on the other hand, constitute a sort of hippy commune, happy and feckless.*

At least thirteen different examples of tool use in chimpanzees have been recorded, showing a wide range of implemental activity. One of the most interesting is the so-called termite-fishing behavior which was first observed by Jane Goodall in 1961 among a group of animals she was studying in the Gombe Stream National Park in Tanzania. At certain seasons of the year the chimpanzees go fishing—fishing for termites that is, poking a long, slender twig or vine down the flight holes, through which the winged termites emerge to engage in their nuptial flights, into the heart of the nest. When the chimpanzee withdraws the twig, it is well covered with worker termites clinging on by their mandibles. The termites are then picked off by the lips and tongue (Fig. 43).

The most interesting part of this activity is the preliminary collection of suitable twigs from areas as much as 15 yards from the termite nests. Considerable care is exerted in choosing "fishing rods" of suitable length, diameter, and flexibility. When chimpanzees settle down for a termite orgy they have a number of spare rods ready for use.

The action that elevates termite fishing from mere tool-using to tool-modifying is the careful pruning by the chimpanzee of leaves and side-branches that would impede their use as probes. If we label this sort of activity a crude form of tool-making, we make nonsense of the definition of humans-the-tool-makers. So, in order to restore the perspective, we must—as Louis Leakey once said—either redefine tool-making or redefine man. In this book I have preferred to redefine tool-making by creating a third category of tool-modifying within the evolutionary stream of tool-using to tool-making.

As far as the status of chimpanzees as tool-users is concerned, we must recall that the chimpanzee hand is anatomically ill-equipped for tool-making. A shortness of the thumb in relation to the great length of the fingers largely accounts for this. There are other factors too that militate against an effective power and

* See also Tuttle, *Apes of the World* (1986).

Fig. 43. Gombe chimpanzee "fishing" for termites in a mound. (Courtesy of Jane Goodall)

precision grip. In this sense the chimpanzee has reached its technological ceiling. Sadly, the preparation of a suitable rod for termite-fishing is about as far as it can go.

Already we are falling into an anthropomorphic trap. "Poor chimpanzees," we say, "how sad to be so deprived!" What possible incentive is there for chimpanzees to make tools? But why should chimpanzees be pitied for their inability to construct a hand-axe, an engraver, bone needles, the spear, and the arrow? They have no need for objects like these.

It is only the extroverted, inquisitive personality and innate intelligence of these apes that, in captivity, make them suckers for anything new. They will romp through a zoo "tea party," pile boxes and join two sticks to obtain a reward, learn sign- and symbol-language without tears, and generally ham it up to please humans with unfailing enthusiasm.

The Osteodontokeratic Culture

Nevertheless the superiority of chimpanzees in tool-using over any other nonhuman primate receives recognition by erecting a special category of tool use called tool-modifying. As tool-modifying is a grade within the continuous process of biological improvement, early hominids must inevitably have passed through just such a stage on their way to full-fledged tool-making. Perhaps the much maligned osteodontokeratic (bone, tooth, and horn) culture of Dart is just such a case; it has a lot in common in terms of the complexity of activity with the inspired ingenuities of Jane Goodall's Gombe chimpanzees.

In a memoir published in 1957, Professor Raymond Dart described the vast accumulation of animal bones, teeth, and horns discovered at Makapansgat, mostly (92 percent) bovid in origin. He argued on statistical grounds that the evidence of selection of the material (as opposed to a chance accumulation) could not be explained simply in terms of animal predators and scavengers. Instead Dart proposed that it was the work of the gracile australopithecines, themselves hunters and animal-killers.

The selectivity of the accumulations, which was Dart's main argument, is exemplified by an over-representation of lower jaws, cannon bones, the distal portion of the humerus, and a large number of bony flakes derived from spiral fractures of long bones thought to be deliberately produced. Other parts of the postcranial skeleton were under-represented One criticism leveled at selectivity is that there is a striking similarity in the parts that survive in different sites where bones have, unequivocally, accumulated by natural means. Dr. C. K. Brain reports on the bones of goats fed on by Hottentots and their dogs in the Namib, and which were then discarded to lie on the desert surface around the Hottentot villages. Calculations of the percentage survival of each part were made and when the results were compared with the percentage survival found by Dart at Makapansgat, the similarity is striking. Brain's view is that, like the natural accumulations of goat bones, the contents of the

Makapansgat accumulations are entirely predictable and dependent on the intrinsic strength characteristics of each bone. He concludes that no tool-orientated selectivity need be inferred, and with that the main prop of Dart's argument appears to fall to the ground.

The Makapansgat accumulations are found imbedded in the solidified breccia of limestone caves and the question arises as to who or what was responsible for bringing them there in the first place. If australopithecines were not the chief agency, who or what was? Porcupines are prime suspects, as they are inveterate "bone-collectors"; the other suspect, the hyena, is not a collector; it feeds out on the savannah where the food is. This leaves carnivores such as leopards or sabre-tooths that, according to Dr. C. K. Brain of the Transvaal Museum, predated australopithecines. Even if leopards were not the only agency, they—at least—may have been responsible for the punctured and crushed australopithecine skulls, particularly of juveniles. It is easier to accept this interpretation than go along with Dart's belief that australopithecines were killers given to internecine murder and cannibalism.

Professor Phillip Tobias of the University of the Witwatersrand has over the years studied Dart's evidence most carefully. At one time he attributed the collections entirely to human activity (without homicidal overtones), but now he is inclined to take the view that a multiplicity of agencies were responsible—humans, hyenas, porcupines, leopards, and sabretooths. As Tobias has said to me, there is a strong *prima facie* case to expect that australopithecines would have at least as much implemental potentiality as the great apes, if not more. We have already seen that chimpanzees, or at least some groups anyway, are tool-modifiers. Among the osteodontokeratic tool kit exhaustively analyzed by Dart there are plenty of examples of modification of "a naturally occurring object to improve its performance": horn-cores, which are used as wedges to split open the shafts of long bones; the modified humerus with its proximal extremity removed, to be used as a bludgeon—the characteristic weapon of the gracile australopithecines; and even "scoops" made from cannon bones designed to make life easier for edentulous australopithecines! There is sufficient evidence among the animal bones in the accumulations that at least some of them were the work of *Australopithecus*.

Abbe Breuil, the noted archeologist, in a discussion of the bone accumulations at Choukoutien,* near Peking,** stated that "a Bone Age preceded a Stone Age." This may well be so, as the phylogeny of implemental use (from tool-using to tool-making) suggests, but there is also strong evidence from different parts of the world that bone industries continued to flourish parallel with stone industries throughout the Pleistocene (Fig. 44). For instance, the cave sites of Peking *Homo erectus* at Choukoutien showed evidence of broken bones being used as implements. Indeed the modification of bone and antler has continued to this day. Still to be picked up in antique shops are nineteenth-century "apple-corers" made from the cannon bones of sheep. These bony bygones, much used in farmhouse kitchens, are almost identical with scoops found at Makapansgat and identified by Dart as feeding aids for toothless australopithecines.

Stone Cultures of the Plio-Pleistocene

The seeds of doubt that humankind was more ancient than the Last Flood began to pervade scientific thought in the 1830s when flint tools were found associated with the bones of antediluvian animals in cave deposits in Belgium, in Kent's cavern, Torquay, at Abbeville, in the ancient gravels of the Somme, and, later, at St. Acheul, near Amiens. In the same year as Darwin's *On the Origin of Species* was published, Joseph Prestwich, the eminent geologist, gave a paper to the Royal Society, freely accepting the evidence that humans were not postcatastrophic newcomers but ancient inhabitants of the Western world. Where there were stone tools, there had been people; so the hunt for what Boucher de Perthes called "these rude stones" was on. They came to light in their thousands from all epochs, pur-

* Now Zhoukoudian.

** Now Beijing.

ABSOLUTE AGE BP	AFRICAN AND EUROPEAN STONE INDUSTRIES			GEOLOGICAL SUB-EPOCHS		ABSOLUTE AGE BP
10,000	Iron, Bronze, Neolithic & Mesolithic industries			HOLOCENE		10,000
50,000	Blade-tool industries (e.g. Aurignacian)		Middle & Upper Blaeolithic	UPPER PLEISTOCENE		50,000
150,000	Mousterian Tayacian	Hand-axe industries (e.g. Acheulian)				150,000
250,000		Flake industries (e.g. Clactonian)		MIDDLE PLEISTOCENE	LATE	250,000
350,000					EARLY	350,000
450,000						450,000
550,000						550,000
650,000	African only					650,000
750,000						750,000

850,000	Pebble-tool industries (e.g. Oldowan)	Lower Potaeolithic	LOWER PLEISTOCENE	850,000
950,000				950,000
1,050,000				1,050,000
1,150,000				1,150,000
1,250,000				1,250,000
1,350,000				1,350,000
1,450,000				1,450,000
1,550,000				1,550,000
2,000,000				2,000,000

Fig. 44. A correlative scheme bringing together lithic and geological subdivisions of the Pleistocene.

portedly even as far back as the Eocene, the dawn era of primates! Some specimens of course were legitimate and easily recognizable tools such as Acheulian hand-axes and other well-established tool types from the Pleistocene, but the majority (including the Eocene examples, which were originally called eoliths) have not received universal acclaim, being regarded (quite rightly as we now know) as accidents of nature instead of human artifacts. Eoliths are products of natural agencies resulting from river or wave action, frost and soil creep, in which powerful friction of one stone against another produced flakes that mimicked such well-attested stone implements as blade tools, Mousterian scrapers, and pebble-choppers.

There is no reason why eoliths should not have been used on an ad hoc basis by early humans. Oakley comments that naturally occurring flakes are used by some Australian aborigines to this day to fashion wooden objects. It is possible too that some eoliths are indeed the work of humans, but how is it possible to distinguish the artifacts from the natural objects? As a French archeologist talking about eoliths remarked: "Man made one, God made ten thousand—God help the man who tries to see the one in the ten thousand."

The earliest undoubted standardized artifact is the pebble tool, or pebble-chopper as it is now called, that dates back at least to the end of the Pliocene, 1.87 million years BP, at Koobi Fora, Kenya, and probably earlier still at Kada Gona, Hadar, Ethiopia (2.5 million years BP). The best known site is Olduvai Gorge where, for the first time, living sites* were discovered with occupants, food bones, and implements side by side. The potassium-argon dating technique puts the time of the earliest living sites at 1.9 to 1.75 million years BP. Pebble-choppers (Fig. 45) are made out of cobbles from which a number of flakes have been crudely struck off from both faces to form a cutting edge where the flake scars meet. The butt of the implement that is not worked is smooth and rounded and fits comfortably in the hand. Other tools such as spheroids and flakes were present in the

* This interpretation has been challenged, for example, by Binford (1987) and Potts (1984).

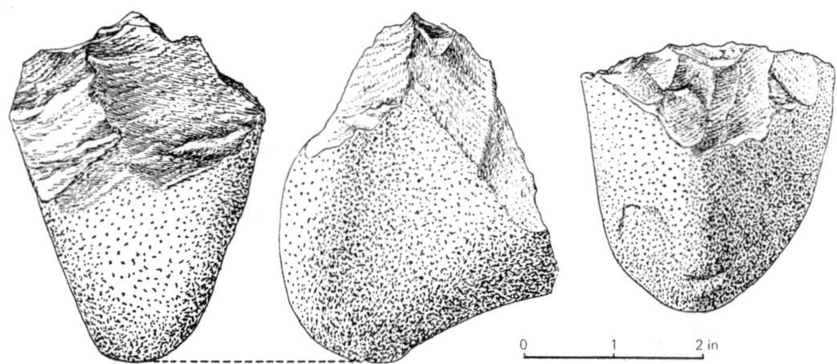

Fig. 45. The pebble-chopper—forerunner of the hand-axe, represents the most primitive form of fabricated stone tool. Pebble-choppers were in use for at least one and one-half million years. (Courtesy of the Trustees of the British Museum [Natural History])

same deposits but undoubtedly the pebble-chopper was the principal tool and was used in a power grip (Fig. 46b).

To the untrained eye, pebble-choppers look like naturally flaked stones, but there are at least two compelling reasons why they are unquestionably artifacts. First, a similar flaking technique is repeated with remarkable consistency on thousands of specimens showing, in Leakey's words, "a set and regular pattern"; secondly, the material used does not always occur naturally at the site. The raw material used at Olduvai includes lava and quartzite, which must have been conveyed there by human agency. Such introduced materials are called "manuports" by Mary Leakey. This type of experimental technique was developed in the first place by Dr. Grover Krantz, the expert on the footprints of the Sasquatch, or "Bigfoot," of North America.

The crudest method of preparing a pebble-chopper is by the stone-on-stone technique. The hammerstone is held in one hand and the pebble in the other, or it can be rested on the knee (Fig. 47). It is not easy using this method to exert any fine control. Even the earliest Acheulian hand-axes were made by this technique.

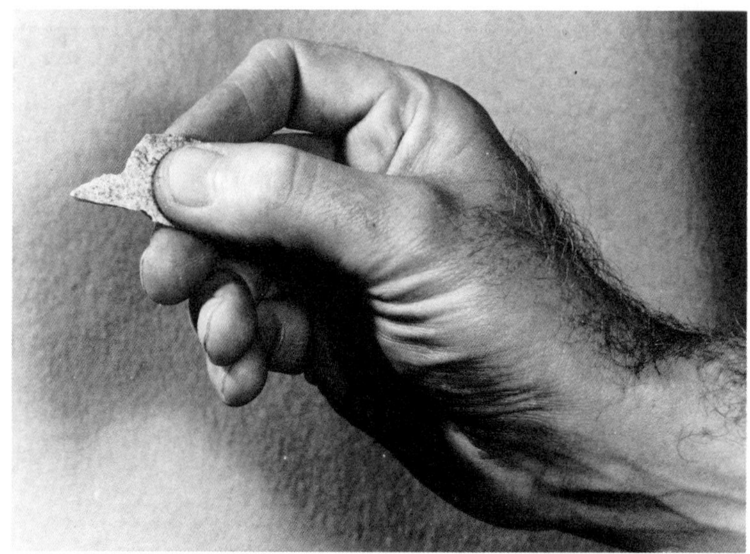

(a)

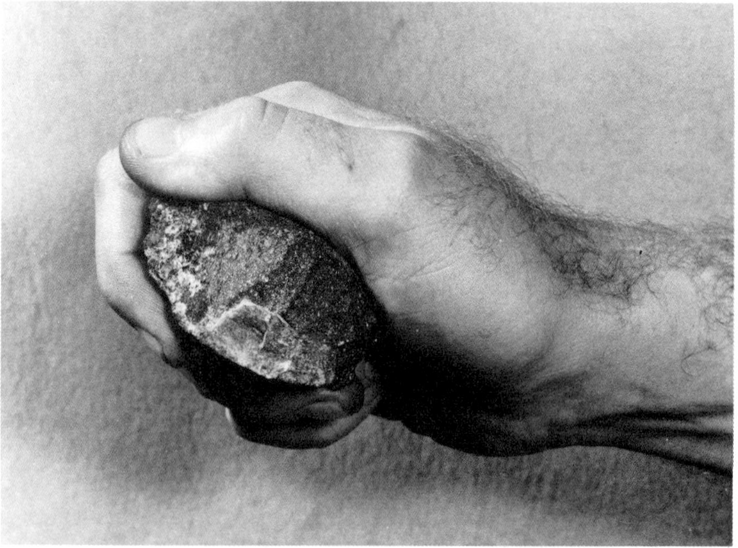

(b)

Fig. 46. (a) Aurignacian burin held in a precision grip
(b) Pebble-chopper held in a power grip

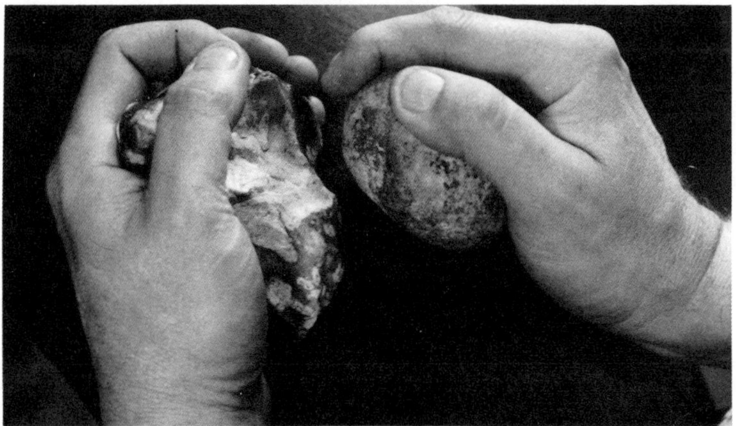

Fig. 47. Author attempting to construct a hand-axe in flint by stone-on-stone method, using power grip.

The pebble-chopper, the characteristic implement of the Oldowan culture, persisted well into the Middle Pleistocene, but hot on its heels was its derivative—the hand-axe. Pebble tools were still widely employed when the first hand-axe was made in Africa at least 1.5 million years ago and they continued to develop side by side. It is generally agreed that the hand-axe was an extension of the pebble-chopper. Hand-axes, to begin with, were made by the stone-on-stone method and the resultant flakes were crude and deep-biting; by alternating the flaking process from one surface to the other, a sinuous or zig-zag cutting edge was achieved. As hand-axes continued to improve over the next 1.25 million years, they assumed a more oval or pear shape (Fig. 48). Flaking was delicate and shallow and, consequently, the cutting edge became less sinuous. A "hammerstone" in the shape of a wooden baton or an antler replaced stone to be followed by the even more precise technique of pressure-flaking.

Hand-axes were a general purpose tool—a scout-knife—that could cut, scrape, and stab the skins of wild animals preparatory to skinning. They were not hafted but held in the hand in a

Fig. 48. Hand-axes showing refinement and improved technique between Lower Acheulian, 500,000 years ago (top) and Upper Acheulian, 100,000 years ago.

power grip. As with pebble tools, it is perfectly possible to make a crude form of hand-axe using the power grip only (see Fig. 47), thus demonstrating that the "tools are as good as the hands that make them."

The Lower Acheulian in Africa is represented in the fossil remains of *Homo erectus* at Olduvai Gorge in East Africa and Ternifine in Algeria. Outside Africa, advances in stone-tool use among populations of *Homo erectus* were much slower. For example, no artifacts have been found in Java, and at Choukoutien, the home of Peking man, only pebble tools and flakes of a developed Oldowan culture are known; Acheulian hand-axes never appear. In Europe, the earliest recorded hand-axes are from Nice in the south of France, dated 400,000 years ago. The hand-axe influence spread slowly northward, reaching England at Swanscombe, Kent, and Furze Platt in Berkshire by 250,000 years ago. The hand-axe culture persisted in Europe until the first warm interval of the last glacial period.

Flakes have probably always been used whenever any form of stone-knapping was practiced. It would be natural to make use of the debris of sharp-edged slivers of flint, chert, or obsidian even if the construction of, a core-tool was the primary consideration. Flint flakes can be used for all sorts of modern purposes, including shaving and preparing a hare for the pot, as I have proved to my satisfaction. Thin sharp slivers of flint that could be inserted into the joints to sever the intra-articular ligaments were particularly valuable. Even at the site of primitive pebble-chopper industries, some of the flakes provided evidence of use.

It is a matter of surprise to me that a true "flake industry" did not evolve earlier than it did. The Clactonian (circa four hundred thousand years ago) was the cultural period when craftsmen started to specialize in flake tools, converting them into scrapers, knives, and even "spoke-shaves" for shaping wooden spears (Oakley, 1967). The Clactonian people appear to have reversed the hand-axe tradition by keeping the flakes and throwing away the cores, as at Swanscombe's lowest levels, where hand-axes are absent. In the upper gravels of the River Thames at Swanscombe, hand-axes return to the scene as people steeped in the Acheulian tradition supplanted the Clactonian tool-makers. The Clactonian exerted a powerful influence, as it seems very probable that it developed into the Mousterian culture that kept Neanderthalians alive for seventy thousand years (see Fig. 44).

The Upper Paleolithic industries (Chatelperonian, Aurignacian, Gravettian, and Magdalenian in order of succession), collectively known as *blade tool cultures*, represented a vast and very sophisticated leap forward in the social and technological history of humankind. *Homo sapiens* was now on the scene. He had evolved a large brain, small teeth, and dextrous hands, morphologically identical with those of modern humans. The best known of these populations were the Cro-Magnons who had spread from Southwest Asia into Europe, where geographically and culturally they replaced the Neanderthalians. Cro-Magnons exhibited, as Oakley says, "a remarkable aesthetic sense and displayed artistic skills scarcely excelled in any later period." Their use of materials had expanded to include bone, antler, and ivory as well as wood and flint.*

The characteristic implements of the Upper Paleolithic, a far cry from the relatively crude chopper-hand-axe tradition, consisted of delicately wrought blades of stone, which included knives, chisels, and engraving implements called burins. The Cro-Magnons also made harpoons of antler, needles of bone, and "Venus" figures (Gravettian) shaped from limestone; above all they were responsible for those artistic wonders of the ancient world—the Franco-Cantabrian cave-paintings.

The nature of tool-making was a very simple one up to the end of the Mousterian—a matter of choppers, axes, and flakes that had preoccupied people of the Old Stone Age for nearly 2.25 million years—an incredibly long period of technological stagnation. However, the progress of stone technology must be judged against the background of the life-style of the tool-makers themselves.

The most important aspect of the life of Old Stone Age peo-

* See Trinkaus, *The Emergence of Modern Humans* (1989); and Klein, *The Human Career* (1989).

ple was food-getting—hunting or scavenging of meat and the gathering of berries and edible roots. Primitive humans were preoccupied with this need from dawn to dusk and the implements they used reflect this single-mindedness. There was no time in the daily round for such trivia as clothes, utensils, or artistic embellishments to cave walls. These activities only came about when populations were larger, when hunting techniques were more efficient and less time-wasting, and when the discovery of fire served, among its many functions, to prolong the length of the working day. Spare time was a luxury that the Olduvai people, the *erectus* populations, and even the early *sapiens* group could not afford. It was not until the last glacial was virtually over and game was plentiful once more that time could be spared for the more leisurely, less vital activities.

Wild gorillas, unlike chimpanzees, do not use tools; they have plenty of spare time, but lack incentive. Early people, on the other hand, had plenty of incentive; what they lacked was time.

CHAPTER SIX

Handedness

TIBETANS have a saying: "Beware of the devils on your left hand side." This maxim is common to every race and creed in one form or another. In most cultural groups the left hand is regarded as both unclean and unworthy. Among Middle Eastern and North African countries the ritual of eating with the right hand only is strictly adhered to in orthodox circles; an awkward procedure at the best of times, but taken all in all, it is probably an important hygienic one, since it is the left hand that is often used exclusively for toilet purposes. In spite of strict injunctions about washing, the left hand must always be defiled, philosophically if not bacteriologically, by the excremental tasks it is called upon to perform.

In the English language the very words "left" and "right" carry their own commendation or detraction as the case may be. The right implies correctness and propriety, the left, in its primary sense, means weak and worthless. Many derived words like righteous and dextrous, gauche and sinister retain their original etymological significance—skill and goodness versus incompetence and evil. Politics, of course, are a special case, although there are those who might not admit to the distinction. According to H. L. Mencken, the use of the Left and Right in politics stems from the seating of the members of the French Assembly in 1789; the conservatives of noble birth sat on the right of the presiding official, the revolutionary members of the Third Estate to his left, and the moderates in the middle.

In the north country dialects, cack-handed means left-handed. To cack, a word in common use in medieval English, means to defecate. The implication of cack-handed therefore is that the individual, left-handed, employs the hand used dominantly in toilet procedures. One wonders if those in whose vocabulary cack-handed still holds a place, realize that they are using the medieval equivalent of a modern four-letter word. According to Partridge's dictionary, cack has a number of variants in different areas of Britain: cork-handed appears in Derbyshire; gallock-handed in East Yorkshire; bang-handed in the Tyne and Tees area; keggy in Castle Bromwich; wacky in Evesham; spuddy-handed in Gloucestershire; scrammy in Bristol; kefty in Somerset; corrie-pawed or corrie-fisted in different parts of Scotland.

From the dialectal obscurity of Devonshire, Norman Capener, scholar and orthopedic surgeon mentioned more than once in this book, resurrected "couchie"—which he suggested was a corruption of gauche. Barsley in his excellent little book—*The Left-handed Book*—has collected some of the words commonly used in foreign languages showing that the attitude toward left-handedness and left-handers is no less derogatory and disapproving than in Britain.

Handedness, like bipedalism and tool-making, is popularly supposed to be a unique possession of people. However, although it may be true that most animals (with notable exceptions, like the male fiddler-crab with its giant right pincer for example) do not overtly demonstrate any lateral preference, there seems to be no doubt that higher primates, at least, show some degree of handedness.

Ambidexterity means the equally skilled use of both hands, but it should be implicit in the definition that the hands are skilled in the first place. The term therefore should probably not be used except where some degree of skilled handedness is established. A horse can hardly be said to be ambidextrous although one might recall, somewhat irrelevantly, the famous case of Clever Hans, the German Wonder-Horse, which astonished audiences by accurately tapping out a number selected by the audience using either front hoof.

From horses to the drawing of horses, one thinks naturally of Sir Edwin Landseer, one of many artists reputed to be ambidextrous; in his case the evidence that he possessed this talent is unequivocal. History relates that he could draw a portrait of a stag with one hand while simultaneously drawing the likeness of a horse with the other, neither being the less perfect. The ambidextrousness of Leonardo da Vinci is a matter of perennial argument. Certainly Merekowski, his biographer, does not believe him to have been ambidextrous. Capener, an expert on the anatomy and surgery of the hand, has expressed some doubt as to whether Leonardo possessed any inborn left-hand skills at all, over and above those normally possessed by people. It is known that a brush with a gang of muggers left his right hand badly mutilated. He may certainly have trained his left hand to take over some of the duties of draftsmanship from his mangled right. There is no doubt that, short of full-blooded ambidexterity (of the Landseer pattern), which is rather rare, individuals possess different degrees of ambidextrousness and in such cases transfer of tasks from one hand to the other would be a relatively easy progression.

Capener is in the minority when he asserts that Leonardo was right-handed; most authorities support the sinistral theory, including Merekowski who believes that Leonardo drew and wrote with his left hand and painted with his right. Leonardo's mirror-writing in his *Notebooks* is often quoted in support of left-handedness but may be quite irrelevant. He may have used mirror-writing not because he was left-handed but because he wished to keep his private thoughts, some of which were said to be extremely radical, to himself. Barsley suggests that Leonardo was a wayward genius who liked to puzzle others—another possible explanation. Other evidence of left-handedness includes the positioning of the cranks in his mechanical drawings such that they could only be used by the left hand. This evidence, perhaps, is more telling but unfortunately does not indicate which hand he *painted* with. Mixed-handed individuals (as distinct from those truly ambidextrous) may be right-handed writers but left-handed manipulators. The problem of Leonardo's handedness remains an unsolved, mystery.

It has been estimated that in 1980 there were approximately 200 million left-handed people throughout the world. This probably represents about 5 percent of a population of 4,000 million. Such a figure, however, is fairly meaningless for a number of reasons, one of which is that it does not distinguish between the frequency of left-handedness at birth and in adult life; in other words there is no means of knowing whether or not the estimate takes into account infant left-handers who have been forced for various reasons into changing their ways. Dr. Bryng Bryngelson of the University of Minnesota estimates that if there was no interference by parents and teachers "34 out of every 100 children born today would become left-handed." Left-handedness—or rather the toleration of left-handedness—appears to be on the increase. A few years ago 4 percent was considered reasonable within a given sample; nowadays the figure has risen to between 5 and 10 percent. There are obviously many factors that would influence the frequency apart from parental and teacher influence, such as intelligence, mental retardation, race and even national ideology.

It may seem strange to include national ideology as a formative factor in handedness, but consider for a moment the likely attitude of a totalitarian regime toward the mavericks of the state who persist in resisting the precepts of national conformity. One can imagine the dictate to all schools that pupils be taught to employ their *right* hands. Nonconformity among the young, it would be said, provides the breeding ground of dissidence. Without the appropriate statistics, one would suppose that the Communist-dominated countries show the lowest rate for left-handedness (except possibly the People's Republic of China). In the United States, Britain, most Western European countries, and Israel, according to Barsley, considerable tolerance is shown. Perhaps this is true at an overall national level, but it would be interesting to know whether there is a similar tolerance, in Britain for instance, in conservative and elitest institutions. Are there any left-handed Wykehamists? Does Sandhurst permit the southpaw? Or crossing the Atlantic, would the FBI tolerate the sinistral-handed among its agents?

One of the largest series of statistics available is that of Karpinos and Grossman (1953) who analysed the manual preferences of 12,159 male U.S. Army recruits, including rejects as well as successful candidates, with the following results:

R. hand preferences	91.4%
L. hand preferences	8.0%

The investigation was carried out by questionnaire, generally considered to be the fairest method. A significant fact emerged and that was that left-handedness was more frequent among the rejects than among the acceptances. A low IQ has often been regarded as consonant with left-handedness, but a recent study by Newcombe and her colleagues refuted this. Indeed the IQ scores in Newcombe's series tended to be lower among those classed as pure right-handed than among the mixed-handed group. On the other hand, Barsley states that the incidence of left-handedness in mental institutions may be as high as 30 percent.

Handedness in Nonhuman Primates*

The first statistical enquiry into handedness in apes was probably Finch's study of thirty chimpanzees carried out in 1941. The animals were required to obtain a food reward by means of manipulatory tasks. Each animal was given eight hundred trials. The results at 80 percent and 90 percent levels were as follows:

	In over 90% of trials	*In over 80% of trials*
R.-handedness	30%	36.6%
Ambiguous	40%	16.8%
L.-handedness	30%	46.6%

The figures reveal, first, that chimpanzees show well-marked handedness (60 percent in the first column above and 82 percent in the second column, discounting "ambiguous" results). Lateral preference is not clear-cut but if anything there is a slight preference for the left.

Helmut Albrecht and Sinclair Dunnett's West African chimpanzee series was even smaller, but the point of special interest is that it provides the only figures available for wild chimpanzees.**

* For more recent information, see MacNeilage *et al.* (1987); Hopkins and Morris (1993); and Fagot and Vauclair (1991).

** See Nishida and Hiraiwa (1982) and Boesch (1991) for additional studies on wild chimpanzee handedness.

Handedness, as in Finch's study, was clear-cut in ten out of thirteen animals. Lateral preference, however, was equivocal, four being right-handed and six left-handed. The task that Albrecht and Dunnett observed was grapefruit-peeling; the fruit was held in one hand, and the contralateral thumb was used to strip off the skin of the fruit.

Most of the experimental work on monkeys has been carried out in macaques (*Macaca mulatta*—the rhesus, and *M. fuscata*—the Japanese macaque). Warren in 1953 performed tests for handedness in eighty-four rhesus monkeys. The results of Warren's study were as follows:

	In over 90% of trials	*In over 80% of trials*
R.-handedness	27%	31%
Ambiguous	46%	35%
L.-handedness	27%	34%

These figures show handedness is present in macaque monkeys but less strongly so than in chimpanzees (see above). For example at the 80 percent level in chimpanzees, handedness is present in 82 percent while at the same level in macaques the figure is 60 percent. A weak left-handed preference is shown, which is probably not significant.

A number of recent studies have broadly confirmed Warren's findings but, with Japanese macaques at any rate, there is a significant preference for the left hand. Tokuda's investigation of forty-one free ranging monkeys gave the following distribution of preference:

R.-handedness	20%
L.-handedness	39%

Comparison between the results obtained by different investigators is complicated by the nature of the tasks used to determine lateral preference since the trend toward left-handedness diminishes as the tasks increase in difficulty. The experiments with free-ranging Japanese monkeys suggest considerably stronger left-hand preferences to those macaque monkeys tested in controlled laboratory conditions.

The Origins of Human Right-Handedness

Modern humans, when all the cultural forces surrounding us are taken into account, are indisputably right-handed. Such monkeys and apes as have been studied show a left-hand bias. The bias is stronger in the case of macaques than chimpanzees. Clearly, therefore, there has been a major revolution in handedness during the course of primate evolution from monkeys to apes, and apes to people. The evidence for lateral preference in early humans is pretty slim and stems largely from studies of the characteristics of tools* and the hints supplied by cave-paintings; thus guesswork is rife. As Thomas Carlyle wrote: "Why [the right hand] was chosen is a question not to be settled, not worth asking except as a kind of riddle." Speculation is the name of the game.

Apart from rather doubtful evidence that a number of fossil baboon skulls from South Africa, contemporary with *Australopithecus*, show signs of a blow from a right-handed adversary (protohuman), most of the evidence of handedness is derived from large collections of human stone artifacts. Oakley reports that the majority of tools from Choukoutien, the site of discovery of the representatives of *Homo erectus*, were chipped by right-handed persons. Sarasin (1918) studied Mousterian scrapers and points and found that there was near-equality in lateral preference:

R.-handedness	34%
L.-handedness	32%

However there are a number of flaws in his work, which become apparent when the function of the tools is taken into account; Sarasin based right- and left-handedness on the manner of grasping alone. Posnansky was more concerned with the functional aspect—the use to which the tools were put. His work was done on a collection of 118 hand-axes from Furze Platt near Maidenhead (now housed at Wollaton Park Museum, Nottingham). By assessing the symmetry or asymmetry of the hand-

* See Toth (1985).

axe, particularly the positioning of the median ridge marking the thickest part of the axe, Posnansky was able to sort the left-hand from the right-hand axes. He also isolated a third group where the hand-axe was more or less symmetrical and could have been used by either hand. When tabulated, Posnansky's results are as follows:

R.-handedness	31.3%
Used by either hand	44.9%
L.-handedness	17.0%
Equivocal	6.8%

Thus, this study points strongly to a right-handed bias in the Lower Paleolithic at least two hundred thousand years ago, but there was still a long way to go before it becomes an overwhelming preference. The emphasis on right-handedness is evident when specialization in such tools as the sickles and the scythes made it desirable that—for pragmatic reasons—there should be a one-handed orientation of society. Agricultural tools were the possessions of the community, not of the individual, and evolutionary pressures would have operated strongly on the genetic basis of handedness in keeping with the community norm. Be right-handed or die of starvation.

The final surge that turned a right-handed bias into the right-handed dominance of modern humans is quite unknown. Possibly it was the pressure of community tools, perhaps the moral force of religio-mystical practices, or the influence of "military training," which taught the men the need to protect the left side of the body, where the heart lies, in hand-to-hand combat with penetrating weapons. Perhaps it was the women who found it more advantageous when nursing a baby in the crook of the left arm to continue their chores with a free right hand; according to Desmond Morris, 80 percent of all mothers cradle the baby on the left-hand side. This figure is confirmed by a recent survey of paintings of the Madonna which estimates that in 83 percent of the paintings the left arm is doing the cradling. This pattern occurs equally frequently among left-handed mothers. Thus it is possible that purely practical considerations, from the mother's point of view, are not as important as I have suggested above. The regular beat of the heart (heard on the left) has been proposed as the important factor. The human heartbeat is already a familiar sound to the infant at birth.

Depictions of the human hand in cave art have been used to "side" the dominant hand among Upper Paleolithic populations (circa 22,000 to 28,000 years ago). According to the Abbé Breuil, hands constitute the earliest examples of cave art at Gargas and El Castillo and other sites in the Franco-Cantabrian province. At Baume-Latrone there are pictures of hands, mainly left, printed with the finger dipped in red clay. At Gargas the technique is different: the depictions are silhouettes outlined in black manganese oxide or in red ochre (iron oxide); these are "negatives." A hand dipped in paint and used as a stamp is a "positive." Negatives are usually of the left hand and positives of the right. At El Castillo there are forty-five negative hands, of which thirty-five are left and nine right. If one assumes that the dominant hand was manipulating the red ochre there was a dextral dominance of 5:1. Contrary findings of the frequency of right profiles of animals over left profiles seem to suggest to Walter Sorell a left-hand bias. He is assuming that the cave-painters had no model and were drawing from memory, which is probably correct as far as it goes, but it is a little dangerous to assume that left-handers always draw right profiles and vice versa.

A unique feature of the stencils of hands at the Gargas cave is that in many cases the terminal bones of fingers are missing. The custom of cutting off parts of the fingers as a sacrifice or a sign of mourning is prevalent in many primitive societies. Since these mutilated hands are always left hands, early humans must have been right-handed as they would hardly mutilate the hand that they depended on for daily existence. It is impossible to draw any scientifically valid conclusions about hand preference from the sort of circumstantial evidence that is derived from cave-paintings and stone tools.

The most that can be said about the origin of right-hand dominance in people is that the left-hand bias of nonhuman primates* appears to have been converted into a right-hand dominance during human evolution; and that this shift was likely to have been a very slow process that did not manifest itself fully until the life-style of early humans started to include activities that required a high degree of delicacy and precision in their performance. A similar sort of argument has been adduced in favor of the gradual improvement (particularly in terms of the precision grip) of the hand itself during the Pleistocene. Thus it may eventually become apparent that *handedness* and *handiness* are interdependent. Without full *handiness*, a dominant *handedness* is really irrelevant.

* Actually, more recent studies show that this is true only of prosimians during acts of one-handed prehension; it is weaker in monkeys and not apparent in the great apes. Indeed, right-hand preferences for manipulations are more prominent in monkeys versus prosimians and especially in apes, particularly gorillas, versus monkeys (MacNeilage, 1991).

CHAPTER SEVEN

Fingerprints

FINGERPRINTS are the positive impressions of the papillary ridges. The occurrence, development, and function of papillary ridges have already been discussed; here we are concerned with the application of fingerprint techniques in the forensic, archeological, and medical sciences. The science of fingerprint analysis is known as dermatoglyphics*, a term introduced by Professor Harold Cummins in 1926. The more archaic term for the science, dactyloscopy, is still sometimes used.

Scientific specialists who study fingerprints are extraordinarily diverse in their interests. Geneticists study prints as heritable structural traits or characteristics that can be recorded. Physicians study them for their diagnostic value in disease such as celiac disease and certain congenital hormonal conditions such as mongolism, Turner's syndrome, and Klinefelter's syndrome. Archeologists note fingerprints diligently because of their occurrence on ancient clay pots and on the fingers of Egyptian mummies. Finally there are the forensic scientists and fingerprint officers whose concern is the identification and classification of accidental "dabs" found at the scenes of crime.

The first thing to understand is that fingerprints are immutable and persistent. They can be obliterated temporarily, but eventually their original pattern is restored. They are also wholly individual. No instance has yet ever been noted where two separate digits possess the same papillary ridge detail.

Papillary ridges, once established, do not change with age, and they are totally unaffected by environment, though they undergo atrophy with loss of pattern in celiac disease and severe eczema, and in skin deprived (for one reason or another) of its nerve supply they almost completely disappear.

* See Cummins and Midlo (1961).

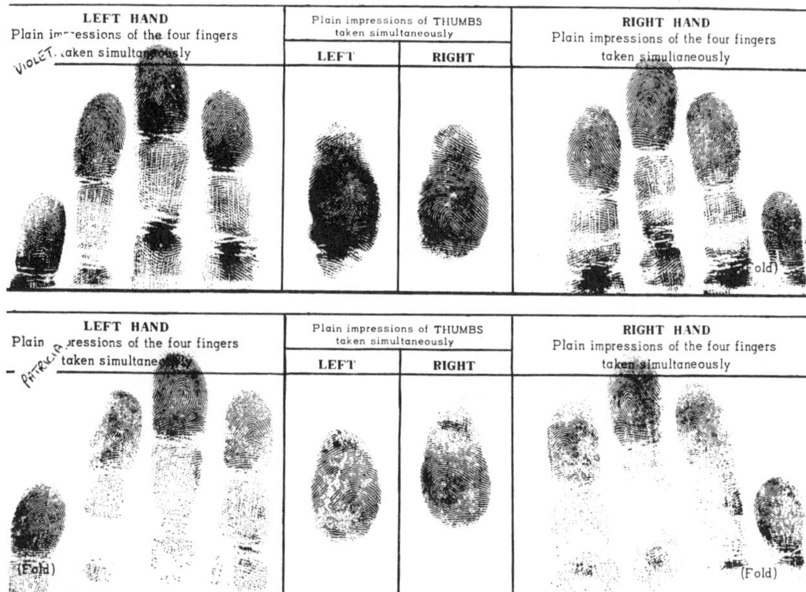

Fig. 49. Police record of a set of fingerprints of identical twins. (Courtesy of the editors of *Fingerprint Whorld*)

Ridges appear in the fetus between the third and fourth months, and it is at this time that the influence of congenital defects of the chromosomes on papillary growth patterns manifests itself. Palmar prints—as opposed to fingerprints—are highly sensitive to such influences, particularly at the site of the atavistic palmar pads at the base of the fingers.

Identical twins are derived from the division of a single fertilized ovum (monozygotic). Nonidentical twins (dizygotic), who have the same chance of looking like one another as any brother and sister, are born from separate ova, each fertilized by a separate spermatozoon. It is a well established belief that identical twins are identical down to the fingertips, but the fact is that although superficially there is a tendency to overall similarity in pattern, their papillary ridge details are quite distinct (Fig. 49). The natural law that no two fingerprints are exactly alike holds for identical twins as for others.

Although the acknowledged inventor of the present-day system of classification and identification was E. R. Henry (later Sir Edward Henry, commissioner of police at Scotland Yard), he was not the first to be aware of the forensic function of fingerprints. Identification by means of fingerprints presupposes the existence of a fingerprint collection for comparative purposes; this is the hard meaning of identification. In the soft sense, identification of fingerprints can also be regarded in the same light as a seal impressed on sealing-wax—recognizable as the trademark of an individual. The earliest examples of fingerprints used as seals date from the third century B.C. when they were quite commonly used by Chinese businessmen.

Prints that have been left accidentally in clay before firing or have been used to embellish the pottery with a design such as a pastry-cook might employ to crimp the pastry, are referable to even earlier dates. Nail-marks were also used by potters as ornamentation and they are to be seen on the rough surface of Assyrian bricks.

In more recent times there are numerous examples of finger or thumb impressions used for authenticating documents. Thomas Bewick (1753–1828), the well-known engraver and miniaturist of natural history subjects, "signed" his works with a fingerprint and the words "Thomas Bewick, his mark."

After the Orange War the citizens of Derry petitioned King William in London for damages suffered during the siege; the document was authenticated by the thumb and fingerprints of the local worthies and (as if to prove that they were well educated and respectable men in spite of the crudeness of their marks) they also appended their signatures.

The awareness of pattern in fingerprints seems to have first wakened in the breast of Malpighi (1628–1694) whom anatomists will know from his "corpuscles" and his "layer." There is the story of an apocryphal food-chemist who in America invented a series of soft drinks called 1-Up, 2-Up, and so on, all of which failed. When 6-Up failed too, he gave up his quest on the brink of success. Malpighi also missed immortality by a hair's breadth, and instead it was Purkinje (1787–1869) of the University of Breslau who is credited with the first classification of fingerprint patterns.

The true inventors of the modern police method of classification are Sir William Herschel, Sir Francis Galton and—the only policeman among them—Sir Edward Henry. The popular system of identification in police work at the end of the nineteenth century was Bertillon's "Anthropometric System," which included *inter alia* the recording of fingerprints, but in 1901 the clumsy and subjective elements of Bertillon's method were dropped in the United Kingdom in favor of a system based entirely on fingerprints. In 1901, the first Finger Print Bureau at Scotland Yard was inaugurated. The first conviction in Britain employing fingerprint evidence was in 1902. After that there was been a steady advance in the use of this new weapon in crime prevention: the first case of murder that rested on fingerprint evidence (1905), the development of a telegraphic code for fingerprint formulae (1921), the single fingerprint system (1930), the use of palm prints (1931). The first murder conviction on the basis of palm prints took place in 1942; the first fingerprinting of a whole town (Blackburn) in a murder case occurred in 1948 (the culprit was eventually caught and hanged).

Acceptance by law of the evidence of fingerprints came in 1948, when the Criminal Justices Act was passed. Section 39 accepted that fingerprints "shall be sufficient proof of a person's previous conviction."

One of the unsung pioneers of dactylography was the Scottish doctor, Henry Faulds (1843–1930). Quite independently, while a medical missionary in Japan, Faulds came to the conclusion that the ridges and patterns showed a high degree of variability and were themselves immutable. In this he foreshadowed the work of Herschel, although it was Herschel who subsequently received much of the credit.

On 28 October 1880 Faulds wrote a letter to *Nature* that was (with the exception of Dr. Nehemiah Grew's treatise in 1684) the first paper ever published on fingerprints. In it, he recounts how his fingerprinting method actually succeeded in solving a crime. An incident took place near his home in Tokyo during which a thief, climbing over a whitewashed wall, left a set of sooty fingerprints behind. A suspect was arrested. Faulds took his fingerprints and compared them with those at the scene of the crime. They did not match and the suspect was released, thus showing how fingerprints may protect the innocent as well as bring the villain to justice. Later the real offender was apprehended and this time Faulds was able to demonstrate a perfect match.

In spite of this and other successful police cases, Faulds did not have much success with the British police when he came back to England. In fact they regarded him as a bit of a crank and preferred to adopt Bertillon's system rather than the ten-finger method that Faulds offered to set up for them at his own expense. Whatever Faulds may have thought of the police, he would no doubt have liked to have known that J. Edgar Hoover, the prestigious director of the FBI, was an ardent admirer of his work and methods.

The anatomy of the papillary ridges is determined partly by "nature" (the heritable component) and partly by "nurture" (the environment). Simple pressure is clearly not conducive to the production of dermatoglyphics. The human heel pad, which bears the greatest weight, never shows papillary ridges, nor do callouses on the hands of heavy manual workers. However, the baboons and others, like macaques and mangabeys, that walk in a *digitigrade* fashion, bearing the weight on the palmar pads at the base of the fingers, show well-developed whorls (Fig. 18), so the matter is not quite so simple as it may appear at first sight.

Ridge configurations occur not only on the fingertips but also elsewhere in the hand, notably the sites of the palmar pads of mammals when a "delta" or triradius (the meeting point of three systems of ridges) normally occurs. On the thenar eminence, ridges tend to be longitudinally aligned; on the hypothenar eminence, they are transverse or oblique.

Fingertip Patterns

According to the Galton classification, fingertip prints are of three kinds: arches, loops, and whorls. Henry added a fourth—tents.

ARCH: Ridges run from side to side in a symmetrical arch-like fashion. True triradii are absent. Arches are rare and constitute approximately 0.5 percent of the population (Fig. 50a).

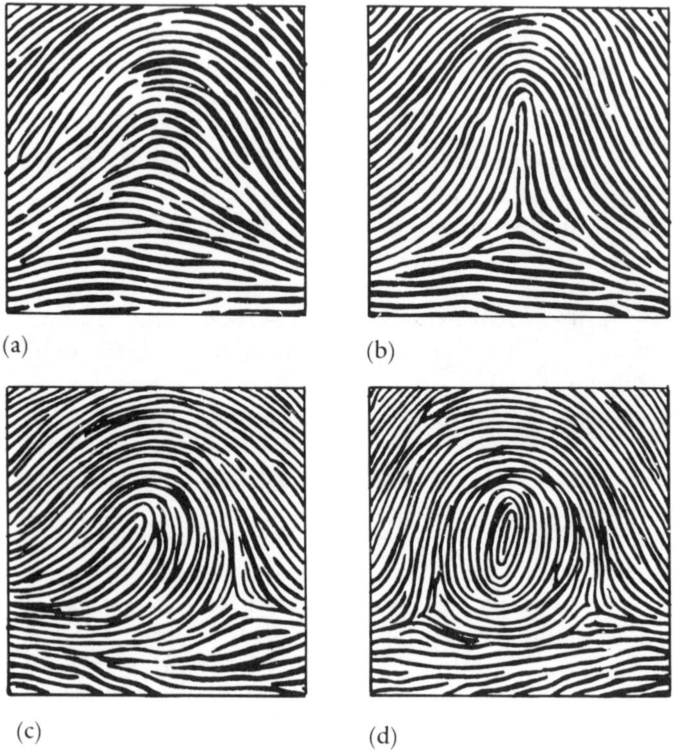

(a) (b) (c) (d)

Fig. 50. The four main fingerprint types:
(a) arch
(b) tented arch
(c) loop
(d) whorl
Adapted from Dr. Sarah Holt in *Genetics of Dermal Ridges*. (Courtesy of Charles C. Thomas)

TENT: This pattern, a variant of the former, produces a vertical upthrust with ridges arching symmetrically over it (Fig. 50b).

LOOP: Loops are of two main types: those that are directed toward the ulnar side and those directed toward the radial; the delta lies on the opposite side to the direction of the loop. Loops are formed of ridges arching sharply round a core (Fig. 50c); in some loops the arching ridges pass so close together that they give the appearance of a whorl. Loops are the most common configuration, constituting 63.5 percent in males and 65.6 percent in females. Of these ulnar loops are far more frequent than radial loops. Loops are seen not only on the fingertips but also on the sites of the ancient interdigital palmar pads, and on the thenar and hypothenar pads.

WHORL: This pattern has a well formed circular or elliptical core with a delta on both sides. One variation of the whorl pattern is the *central pocket* which is essentially a loop pattern in which a few ridges around the core are closed (Fig. 50d). Whorls are second only to loops in frequency of occurrence—26.1 percent in males and 23.9 percent in females.

Ridge Characteristics

Ridge characteristics provide the basis for fingerprint identification. Ridges do not run smooth and uninterrupted courses. Some end abruptly, others bifurcate, some split to form island-like enclosures, others fork to join up with adjacent lines (Fig. 51). There seems to be no particular pattern to the occurrence of ridge characteristics, except that they are most commonly seen in the region of deltas (triradii) and cores. Seven different types of ridge characteristics are recognized. Sixteen points of similarity of ridge characteristics between two prints are required by the law in Great Britain to uphold claims of identity (Fig. 52). The number of points of similarity required by law outside Great Britain varies from six in India to seventeen in France.

For reasons not at all understood, the patterns of the papillary ridges on the fingers and palm duplicate the sort of patterns seen in nature in a variety of different situations. For example, the formation of desert sand dune systems many miles long seen from the air, or the sand ripples left by the receding tide on a shallow beach, are almost identical, even to the presence of all seven of the recognized ridge characteristics. Other examples include such varied phenomena as the bark of chestnut trees, the backs of male cuttlefish, and the coats of zebras. The "papillary" pattern of the zebra skin bears an even closer analogy with fin-

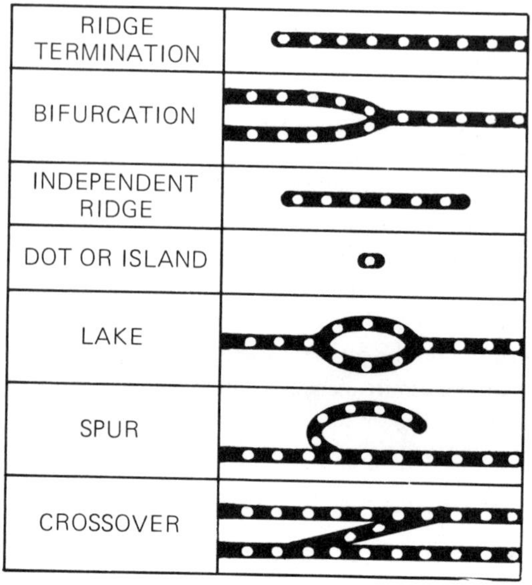

RIDGE TERMINATION
BIFURCATION
INDEPENDENT RIDGE
DOT OR ISLAND
LAKE
SPUR
CROSSOVER

Fig. 51. Variations in papillary ridge systems.

gerprints, which are sometimes seen to bear *subsidiary ridges*. These are extremely fine ridges between two normal ridge prints; being countersunk, they leave an impression only when a strong degree of pressure is used in recording the print on paper. One subspecies of Burchell's Zebra (*Equus burchelli antiquorum*), a southern race, is characterized by "ghost" stripes—narrow, faintly outlined stripes lying between two developed ones.

This duplication in nature of the characteristics of papillary lines is the original observation of two fingerprint experts of the Hertfordshire Constabulary, John Berry and Martin Leadbetter. It is difficult to believe that such a remarkable series of coincidences has occurred haphazardly; instead, one is tempted to look for some fundamental pattern of growth or form.

Ridge Counts

Since the fingerprint types may differ on each finger of individuals and, furthermore, since there is a complete series of intermediate forms separating the four main pattern types, it is impossible to classify individuals according to type. To overcome such difficulties, recourse is made to "ridge-counting," using the method of Bonnevia (1924). The count is based on the number of ridges that are transected by a line drawn between the nearest delta or deltas (triradii) and the center, or core of the pattern. Arches, lacking a delta, have a score of zero. Loops have a single delta and whorls have two, in which case the higher count is used. Ridge counts on all ten digital tips are then added together. Neither terminal—the delta or the core—is included in the count, which may vary from one to sixty. There are rules of procedure as to what constitutes the measuring point of the delta, but it is sufficient here to illustrate a simple example. Subsidiary ridges (see above) are not counted.

The Henry System and its modifications are clearly explained in an HMSO monograph. The author is the famous Chief Superintendent Cherrill, whose name was an honored one in law-abiding homes and a dirty one in every thief's kitchen.

The position regarding racial differences is today precisely as Francis Galton recognized it in 1892 when he compared the fingerprints of English, Welsh, Basques, Jews, and "Negroes." Differences between races are purely statistical and depend mainly on the numbers of loops and whorls in the populations under study. Racial dermatoglyphics is not, and cannot be, concerned with individuals. Such newspaper headlines as "Fingerprints Show Nationality" are rubbish. In a recent analysis involving fingerprint classifications of over half a million individuals of all races, the findings indicate that the frequency of whorls is significantly less in Caucasians than in non-Caucasians; the largest number occurs in Chinese and Japanese. The reverse—as might be expected—is true of loops; Caucasians have the greatest percentage and Mongoloids the least.

According to one authority, the retention or elimination of the palm patterns of the thenar and hypothenar areas is a better

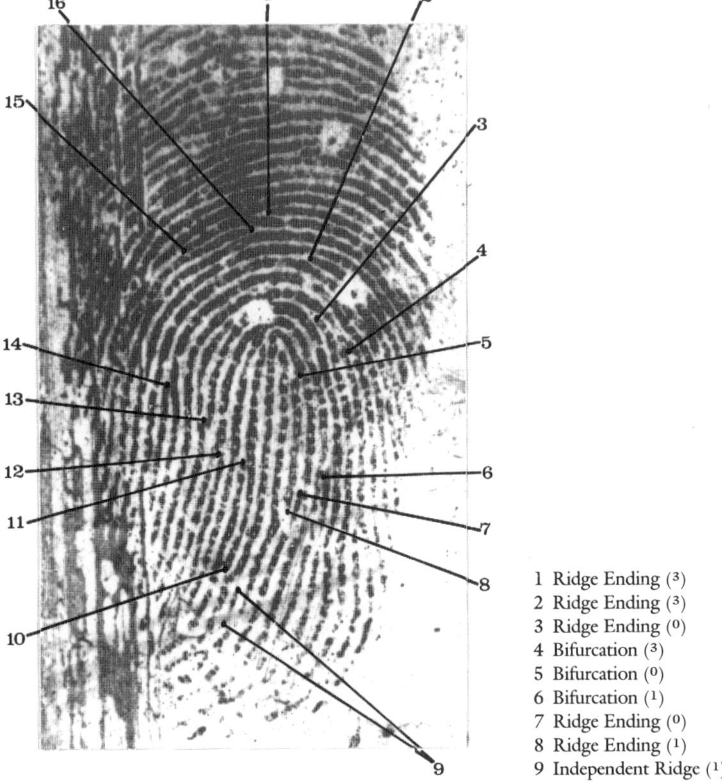

Fig. 52. Characteristics of papillary ridges are used to compare fingerprints using a special formula. In Great Britain the law requires sixteen points of similarity to uphold identity.

indicator of race. The only racial groups displaying more thenar than hypothenar patterns are "Negroes", Mayas, Aymaras, and Kechwas. In all the other races the hypothenar patterns are considerably in excess of the thenar.

Mongolism, or Down's syndrome, is characterized by the presence of an extra pair of chromosomes. The effect of this genetic abnormality is a retardation of growth in most parts of the body,

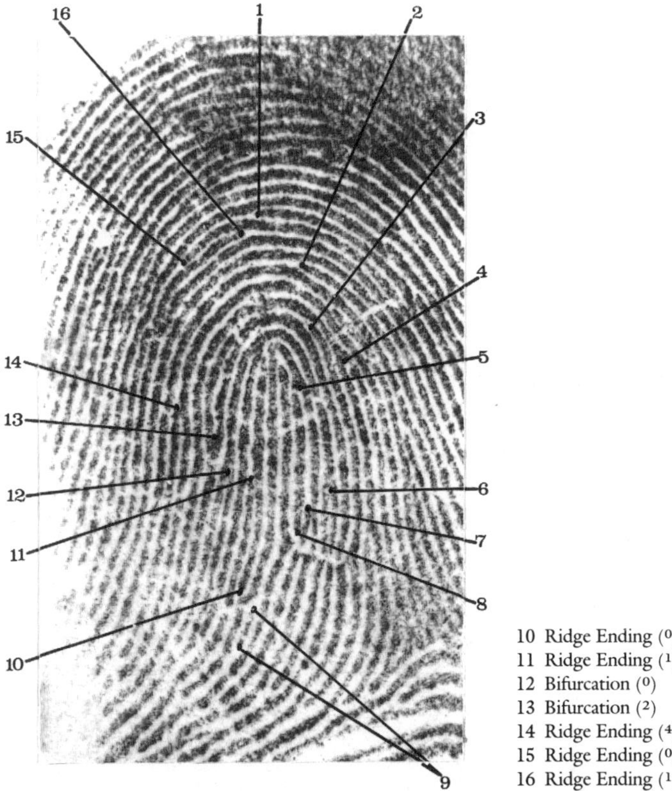

Fig. 52. (*Continued*)

giving rise to multiple abnormalities. Abnormalities are found in the digital and palmar lines of the hand, which, itself, is short, with stubby digits and incurved little fingers. Skin abnormalities include frequently the presence of a simian crease (see p. 33) and the absence of the distal flexure crease of the little finger. The presence of a simian crease has been held to indicate congenital heart disease or rubella in the mother during pregnancy.

Aberrations of the sex chromosomes are associated with conditions such as Turner's syndrome in which ovarian tissue is congenitally absent, and Klinefelter's syndrome, in which *inter alia* the mammary glands are excessively developed in the male. Both these diseases are associated with dermatoglyphic abnormalities.

Atrophy and actual loss of fingerprints occur in a variety of local conditions—peripheral nerve injuries, eczema, and old age, to mention a few. Recently, a relationship between celiac disease, a metabolic condition, and severe atrophy of papillary lines has been demonstrated.

In a fairly recent textbook of comparative anatomy the authors summed up the functional role of papillary lines in an unequivocal and rather sour paragraph: "[Papillary ridges] are useless organs which serve to convict man of animal ancestry as they have on occasion served to convict them of crime." In order to spare these misguided authors more of the remorse that surely must have haunted them ever since, I refrain from mentioning their names. Bearing in mind the role of papillary ridges in manipulation and tactile discrimination, it seems hardly appropriate to call them "useless organs," to say nothing of their genetic, diagnostic, and forensic applications.

As Mark Twain, ahead of his time as he often was, correctly observed in *Life on the Mississippi* and, later, in *Pudd'nhead Wilson*, fingerprints are the only indelible signatures of man. Sorell writes in his interesting book, *The Story of the Human Hand* (1967): "We can fake and forge the written name or . . . we may change the way we sign it. Some people even change their name[s]. Nature permits such freedom of choice, but our heredity rules the signs on our fingertips with authoritarian power for life."

CHAPTER EIGHT

Gestures

HUMAN COMMUNICATION is a complicated business in which all the senses play their part, vision and hearing being dominant. The chief organs of human communication are the eyes, the ears, and the vocal apparatus, a combination that is unique to people. This chapter deals with nonverbal communication.

Gesture forms a notable part of our day-to-day lives. According to Sir Richard Paget (1869–1955), people can make 700,000 different gestures, 150,000 more than the number of words listed in the largest English dictionaries. While the claim is probably extravagant, there is no doubt that sign-language is a very important aspect of human communication.

Like so many aspects of human anatomy and physiology, the use of gestures, their role, and the conditions under which they are used, have been taken very much for granted. As a prime function of the human hand, the use of gesture demands more than a passing comment. The following list is not claimed as definitive, but it comprises the principal contexts in which gestures are employed by people. Gestures include facial expressions and bodily movements:

- As accompaniment to normal speech
- As a substitute for a foreign language
- As a substitute for normal speech
- As a substitute where normal speech becomes inaudible, disadvantageous or dangerous
- As an accompaniment to certain professional activities, *e.g.*, by actors, dancers, and political speakers, to supplement or replace the spoken word

In moderation, gesture enhances the effectiveness of speech; indeed speech without gesture is sterile and tedious. When lack of gesture extends from the hands to the face with a suppression

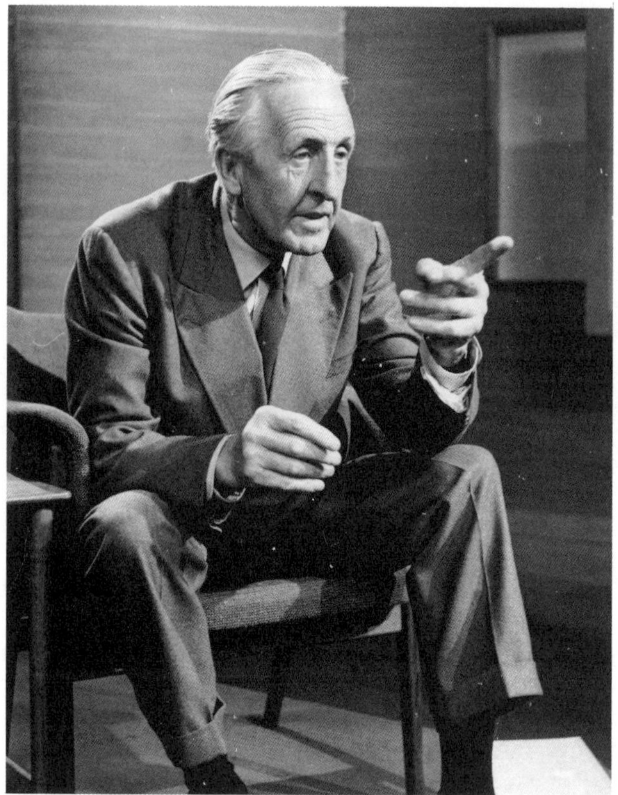

Fig. 53. General Sir Brian Horrocks—a great communicator. (Courtesy of the BBC)

of facial mobility, the general "deadpan" expression is irksome. On the other hand, arms that are waved like the sails of a demented windmill are nothing but an irritating distraction. I remember some years ago a television series of military talks by General Sir Brian Horrocks, a fluent speaker, that were notable for Horrocks' impressive battery of gestures, enthusiastic yet controlled, that were a graphic accompaniment to the dramatic tales of a highly mobile war that he was telling (Fig. 53). "Talking heads," in television jargon, come in all shapes and sizes, and the best are those whose hands and faces are alive, expressing emotions as well as facts. Cold reason and impassive objectivity might be all right for historians but are not the stuff to make compulsive viewing.

While fluent speech is enhanced by appropriate gesture, bumbling and hesitant speech in which speakers have to disinter their thoughts from the dark catacombs of their minds, is often rendered farcical by stereotyped gestures that indicate an agony of mind rather than a fluency of mental imagery. Ethologists would refer to such movements as displacement activities.

Richness of gesture varies with ethnic background. D. Efrom and J. P. Foley studied gestures among Jewish and Italian immigrants to the United States and showed how the extroverted theatrical gestures of the Italians were in contrast to the more intellectual introverted hand movements of the Jews. Sorell regards the use of gesture as owing more to racial origins than to environmental influences. Desmond Morris describes three basic forms of hand-waving: the *vertical*, the "*hidden palm*" wave, and the *lateral* wave. The hidden palm wave indicates that the palm is facing inward while in the other two it is facing outward. He describes the vertical wave as being common in France but rare or absent in Italy or Britain. The hidden palm wave is almost wholly an Italian gesture, and the lateral wave is predominantly British although not uncommon in France.

As a substitute for language difficulties, gestures are obligatory and are often combined with simple words like the immortal, "You Jane, me Tarzan" interchange. Modern sign-languages evolved from the original system invented by Charles Michel in 1759 and taught to deaf mutes at a special school in Paris; nowadays a number of different versions of gestural speech exist, such as the Deaf and Dumb Language and the American Sign Language (ASL). Communication through signs was extensively practiced by the Plains Indians to overcome the variety of languages and dialects among the nations. As far as I know there is no international sign-language today comparable to the spoken lingua franca, pidgin, or creole languages, although the ASL would probably do very well. Morris calls such "languages" coded gestures, a formal system of hand and arm signals. He

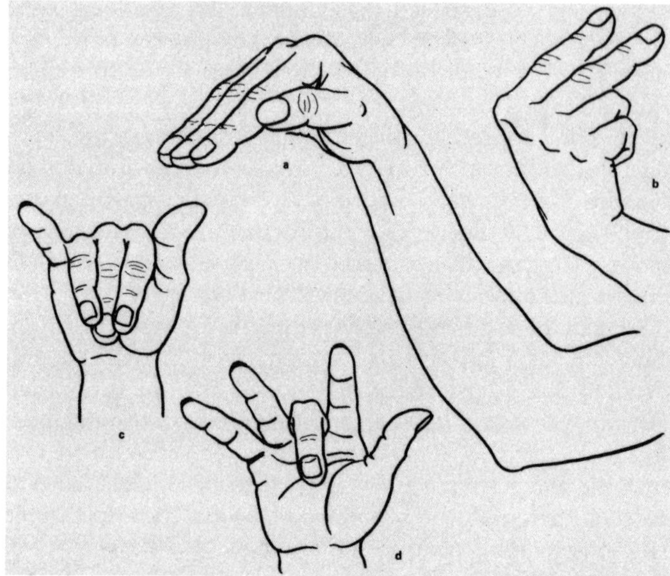

Fig. 54. Hunting gestures used by bushmen to transmit information regarding the nature of the game:

(a) ostrich
(b) eland
(c) bat-eared fox
(d) giraffe
(Courtesy of Irvin DeVore, drawn from photographs)

includes semaphore and the tic-tac system used by bookmakers and their touts.

Sign-languages are also used when a ban on speech is enforced by circumstances—war parties, commando units, and hunting groups. The Kalahari Bushmen have developed a sign-language to communicate with their companions when stalking game; the gestures are pictorial rather than symbolic (Fig. 54), the hands mimicking the overt characteristics of the animal that has been discovered. Certain religious communities who have taken the vow of silence, such as the Trappist monks, employ a sign-language as do those involuntary occupants of cells, the inmates of Her Majesty's prisons. The range of secret signs and handshakes falls into this category. Road-drillers, boiler-makers, shipyard workers, pop-groupers have, no doubt, their own ways of communicating nonverbally against a background of noise.

Cab and lorry drivers, too, have evolved their unofficial sign-language. However, they include some gestures that will never be found in the Highway Code and that are both insulting and effective. The precise posture of the hand in executing a gesture is important, and the distinctions are subtle. Consider the Episcopal blessing, the Churchillian victory sign, and what has come to be known as the "Harvey Smith" gesture. They are all basically two-fingered, but the context in which they are expressed and the precise posture of the hand and forearm varies considerably. The first two-fingered gestures mentioned above are given with the palm facing outwards while an obscene version is performed with the palm facing inwards, often accompanied by an aggressive stabbing or jerking movement; furthermore, the obscenity of the gesture is increased by a reduction of the V-element. The smaller the angle subtended by the two fingers, the greater the insult. Churchill's fingers were always widely spread. His gesture bore no hint of vulgarity but was merely an exhortation to the people for further efforts and sacrifices in the common cause.

Gestures used in professional activities such as acting, mime, ballet, and public speaking possess a ritualistic quality. The actions and gestures are linked to form set-pieces. In the words of John Bulwer, the author of *Chirologia* and *Chironomia* (two books originally published together in 1644):

> the moving and significant extension of the *Hand* is knowne to be so absolutely pertinent to speech, that we together with a speech expect the due motion of the *Hand* to explaine, direct, enforce, apply, apparrell, and to beautifie the words men utter, which would prove naked, unless the cloathing Hands does neatly move to adorn and hide their nakedness, with their comely and ministeriall parts of speech. (*Chironomia*, p. 16)

Bulwer is concerned with rhetorical delivery but, as Joseph (1951) notes, in Elizabethan times, what applied to gesture in

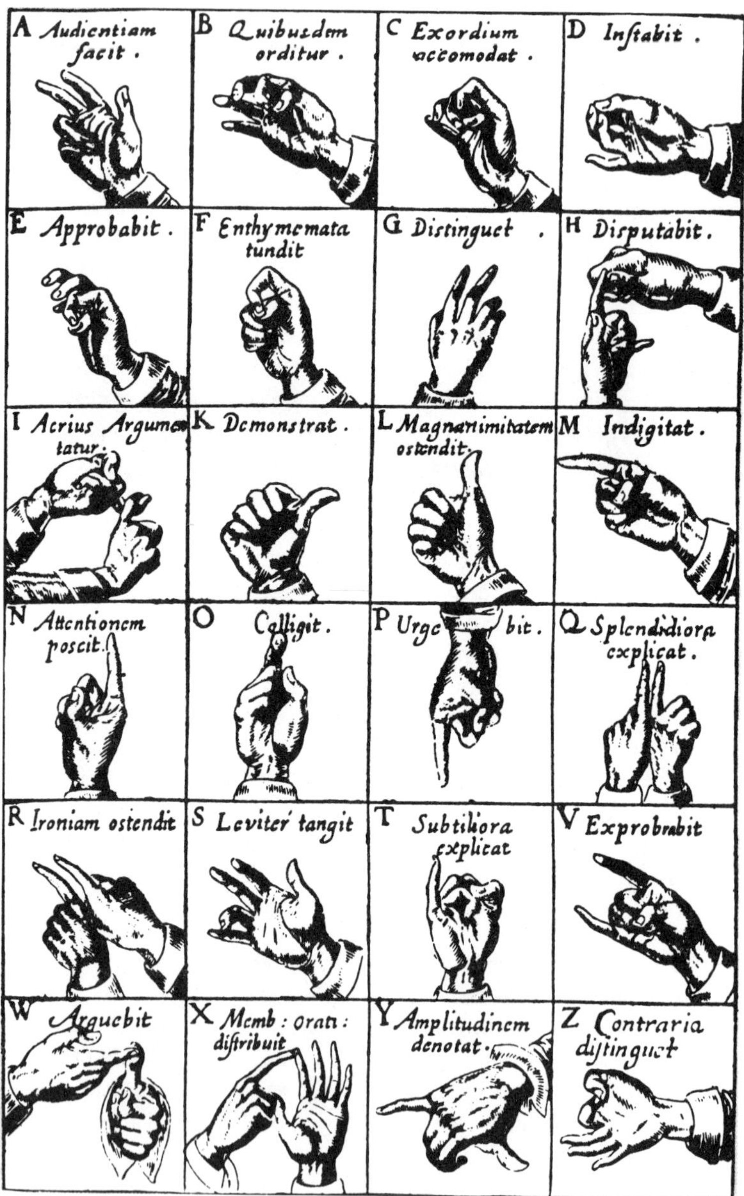

Fig. 55. Classical gestures used in acting and rhetoric in Elizabethan England. (Courtesy of Oxford University Press and Geoffrey Cumberledge, from *Elizabethan Acting* by B. L. Joseph)

oratory also applied to acting on the stage, although Bulwer takes care to distinguish between the miming of the stage player and rhetorical gestures of the orator that flow out of the "liquid current of nature" (Fig. 55).

> Gesture must attend upon every flexion of the voice, not scenic but declaring the sentence and the meaning of our minds, not by demonstration, but by signification . . . to represent a Physician feeling the pulse of the arteries . . . or to show a Lutenist striking the chords of an instrument are kinds of expression to be avoided.

To do so would have smacked of the dancer or actor; constantly he warned against gestures that, in their flamboyance and vehement delivery, belonged more to the theater than to the forum. Joseph notes that in Cicero's *De oratore* the same relation between acting and oratory must have existed in ancient Rome as in Elizabethan England.

If it is true that oratory is dead, then the murderers have been the cinematograph, radio, and television—in that chronological order, the death blow being television. We are perhaps fortunate in Britain in having had at least two great orators in our lifetime—Winston Churchill and David Lloyd George—cast in the classical mold. That at the turn of the century politicians were still striking "rhetorical" poses is attested by the evidence of the early cinematographic records. The coming of radio on the one hand led to a further decline in the graceful art of public speaking and gave way to mere tub-thumping in the marketplace which Bulwer would have assigned in its flamboyance to the theatre and not to the "Churches, courts of commonpleas, and the Councell-Table." The true conventions of the debate in Britain are to be found only in the Union debating societies of Oxbridge.

The coming of television completed the process and in fact brought the art of rhetoric full circle. The modern politician or public figure is placed at a desk or in an armchair, depending on the degree of formality required, arm and hand movements thus constrained and most of the time out of shot. They haven't a chance to irritate the viewer by flinging their arms about and their faces are frequently held in closeup so that extravagant facial expressions are also denied them if they don't want to appear comic or absurd. They are therefore left with the cadences and modulations of their voices and the more subtle expressions of their faces to embellish his oratory. For this art a certain amount of training is required. For all Bulwer's precepts, the television orator has to learn to be an actor. From the elegance of rhetoric, the art of the gentleman, lady, and intellectual, we have turned full circle to the actor who, perhaps, started it all.

Gestures of Dancing

The history of the dance is an even more ancient one. There is some evidence, though speculative, that dancing was practiced in the Late Paleolithic. The Franco-Cantabrian cave paintings show groups of figures in dance-like formations, an observation that suggests that the dance was an element in the religio-mystic ritual that underlies the cave-paintings themselves.

There is a startling difference in the anatomy of the traditional dances of the East and West. The East (India, Southeast Asia) has specialized in somewhat languid dances in which footwork is subordinate to intricate movements of the upper part of the body, especially the head, neck, and hands (Fig. 56). This sort of dance is practiced by people who are slim, flexible, and light-boned, wearing a minimum of clothing (as befits the climate) that does not hinder free movement. It is reputed that Indian dancing comprises at least four thousand positions of the hands and fingers that, when choreographed, tell a story or illustrate a myth. In contrast, the hand in Balinese dances does not tell a story specifically but is used to decorate the movements of the dance as jewels beautify the hand itself. On Samoa and many other Polynesian islands, dances that consist almost purely of hand gestures are characteristic.

The traditional dances of the West, on the other hand, are in every way opposite. They do not favor the upper part of the body; they are energetic, muscular, and often fast and furious, demanding intricate foot-work, such as flamenco, African dances of all kinds, tarantellas, and the wild dances of Southern Europe. In more placid vein and cooler countries clog-dances and Morris dances, including "tap" (derived from clog-

Fig. 56. Expressive gestures of a Balinese dancer.

dancing), tend to display activities that do not lead to undue heat loss.

Males usually play the positive role in Western folk-dancing, women in the East. Compared with the Mudras, the symbolic sign-language of Indian dancing, the mime gestures of Western dance, including ballet, are relatively crude and simplistic; death, life, love, and marriage, are stylized in manual gesture, elegant but unsubtle. Here again the Western preference for "footwork" over "handwork" is apparent.

Gestures in Apes

It is only among the great apes, the possessors of true hands, that gesture is commonly seen and, among them, it is only in the chimpanzee that it is highly developed. A pleading or begging gesture with the hand, palm extended and facing upward, is often performed by macaques. I believe that it is not part of their repertoire in the wild but copied from humans. Temple monkeys in India, provisioned and partly wild Japanese macaques, and Gibraltar "apes" have been observed begging. In zoos, of course, all monkeys have learned this trick, including South American spider monkeys, which "beg" by sticking their prehensile tails through the bars of the cage. The gorilla's use of the hand seems mainly to be tied up in ritualized displays like chest-beating, rather than for casual gestures, as is the chimpanzee's. The orangutan is not known to make any specific gesture but, compared with gorillas and chimpanzees, our knowledge of their natural behavior is rather poor. Chimpanzee gestures are very graphic and remarkably humanoid. There is a dismissive gesture accompanied by a threatening face, a comforting gesture in which the infant or adolescent is gently patted by adult males, greeting gestures in which the hand is held out by the male to receive a kiss from the adult female, and patting-on-the-back and lip-kissing gestures when chimpanzees mutually embrace. Contact-seeking behavior, touching, and embracing bring about much reassurance and comfort. All these gestures and many more have been observed by Jane Goodall in her thirty-year study of wild chimpanzees in Tanzania.*

The hallmarks of humankind are those characters that separate us from our primate relations. There are not many of them. The differences for the most part are quantitative, we being at one end of a spectrum of variation and apes and monkeys at the other. Some characteristics, like tool-making, enjoyed a brief reign as the unique criterion of humanness until Goodall's chimpanzees showed us that even chimps *make* tools (in a modified sort of way), so tool-making is no longer quite the acid test it was once considered. A critical brain size, known as "Keith's Rubicon," was used as the principal touchstone of humanness for many years. Speech and language are now the vogue. Human speech seems to be quite unique—out on its own. As Jane Lan-

* See Goodall (1986, 1990).

caster put it a few years ago: "The more we know about communication (in apes and monkeys) the less these systems seem to help in our understanding of human language."

The essence of human speech lies in a number of unique factors that have been called "design features" by Charles Hockett.* Essentially, such design features allow the communication of abstract ideas, the discussion of events displaced in space and time, the transmission of information through teaching and learning, and the invention of new words. Human speech is not stereotyped like the closed systems of animals such as the chimpanzee, but is open and fluid, constantly ebbing and flowing with changing fashion and the stimulus of human events.

Chimpanzees, because of their overt intelligence as well as their friendly and cooperative temperaments, have often been the subjects of language and speech experiments. Initially, investigations were directed toward teaching chimpanzees to speak, though the projects were singularly unsuccessful. The logical follow-up was to investigate the use of gesture and mime. This has proved a much more successful venture, and communication, albeit at a simplistic level, was established.**

One of the best-known experiments was carried out by Alan and Beatrice Gardner, a husband-and-wife team of psychologists. Washoe, their subject, was a young female adolescent when they started work (Fig. 57). The system they employed was the American Sign Language. The ASL is not a form of finger-spelling, for each position of the hand corresponds to a different word. The Gardners called them "signs"; others have referred to them as "symbols" or "words."

Washoe was taught signs in a number of different ways. One of the ways was to mold a similar and preexisting natural gesture into an acceptable ASL sign. The chimpanzee's begging gesture with palm facing upward was refined into the gesture indicating "Give me." Its derivative, in which a beckoning gesture was added, became "Come." She was taught by placing her hands in

* See Hockett and Altmann (1968).

** See Tuttle (1986).

Fig. 57. Washoe making ASL sign for "toothbrush." (Courtesy of R. A. and B. T. Gardner)

the appropriate positions and rewarding her. Finally she acquired some of her words through observational learning ("tooth-brush," for example). Washoe, like a young child, indulged in a certain amount of nonverbal "babbling," practicing silently the words she had learned.

Occasionally she would invent a new "word." The Gardners could find no sign for "bib" in the ASL manual. They developed a sign for the word but Washoe had ideas of her own and *she* invented her own sign—which in the event turned out to be the correct ASL-ese for "bib."

By all these methods Washoe acquired a rich vocabulary of signs, numbering nearly 100 in all. Her grammar was simple and consisted of two-word phrases like "Washoe sorry," "Please tickle" and "Drink red." Washoe's achievements with two-word combinations are of a similar level to human children's earliest attempts to speak grammatically.

The Gardners tell a story about the meeting of Washoe and a young deaf child at a party. They looked at one another and then solved a socially awkward confrontation in their own unselfconscious way. Washoe rocked her arms, the sign for a baby, and the infant scratched at its armpit, the sign for a monkey.

The widespread use of nonverbal communication by means of manual gesture among human beings and the facility with which chimpanzees learn the skill, in spite of their relatively small brain size, strongly suggests that the combination of vocal calls and sign-language would have served early hominids well enough until speech and language evolved. It has been said by J. S. Weiner, an authority on the Pleistocene, that "it is difficult to believe that co-operative hunting could be performed or skilled tool-making taught without language." Yet hunting, as discussed above, is a silent activity with its own sign-language conventions. What might be more difficult without a symbolic language would be to plan the hunt ahead of time, but why should planning be so important? Individuals would know by experience their roles in the hunting party and would behave accordingly.

As far as tool-making is concerned, demonstration is worth a thousand words, and an observant apprentice to the tribal toolmaker would soon pick up the principles of the technique. Although human in every way—walking, tool-making, hunting—early persons could still have operated effectively without a language. We really do not know at what stage speech (the anatomical component) and language (the grammatical) were established. It could have been at *Australopithecus*, or it could have been as late as the earliest *sapiens* populations of the middle Pleistocene. If brain size is any criterion, then one might favor the time of emergence of *Homo erectus* as the critical period. Between *Australopithecus* and *Homo erectus* the capacity of the brain almost doubled, from 508 cc to 974 cc.

Manual gestures are movements we all make as an integral part of our communication system. Sometimes they serve to replace speech, sometimes to augment and elaborate it, and

Fig. 58. Sketch of hands by Leonardo da Vinci.

sometimes to render it farcical, as did Hamlet's Players, and for which they were properly disciplined: "do not saw the air too much with your hand, thus."

One way or another, gesture constitutes an enrichment of communication and a betterment of understanding between individuals. It is the hidden language that has no vocabulary and no grammar. It allows things to be expressed that can never be spoken of. If language was given to people to conceal their thoughts, then gesture's purpose was to disclose them.

Suggested Reading and References

GENERAL

Bandi, H.-G. 1961. *The Art of the Stone Age*. London: Methuen.

Barsley, M. 1966. *The Left-handed Book*. London: Souvenir Press (also Pan Books, 1969).

Bell, C. 1834. *The Hand; its mechanism and vital endowments as evincing design*. London: William Pickering.

Brandon, J. R., M. Clarke, and H. Koegler. 1975. Dance, The Art of (including Dance, Western and Dance, Eastern). *Encyclopaedia Britannica*, 15th Edition.

Cherrill, F. R. 1954. *The Fingerprint System at Scotland Yard*. London: HMSO.

Goodall, J. (see also Van Lawick-Goodall) 1990. *Through a Window*. Boston: Houghton Mifflin.

Joseph, B. L. (with G. Cumberledge) 1951. *Elizabethan Acting*. Oxford: Oxford University Press.

Leakey, L.S.B. 1933. *Adam's Ancestors*. London: Methuen.

Lewin, R. 1989. *Human Evolution*. 2d ed. Boston: Blackwell Scientific Publications.

Morris, D. 1967. *The Naked Ape*. London: Jonathan Cape.

———. 1977. *Manwatching*. London: Jonathan Cape.

Napier, J. 1962. The evolution of the hand. *Scientific American* 207:56–62.

Napier, J. R., and P. H. Napier. 1985. *The Natural History of Primates*. Cambridge, Mass.: MIT Press.

Oakley, K. P. 1967. *Man the Tool-maker*. London: Trustees of the British Museum (Natural History).

Scott, W. 1967. *The Story of the Human Hand*. London: Weidenfeld & Nicolson.

Tuttle, R. H. 1986. *Apes of the World: Their Social Behavior, Communication, Mentality and Ecology*. Park Ridge, N.J.: Noyes.

Van Lawick-Goodall, J. 1971. *In the Shadow of Man*. London: Collins.

Washburn, S. L. 1960. Tools and human evolution. *Scientific American* 203:63–75.

Wile, I. S. 1934. *Handedness: Right and Left*. Boston: Lothrop, Lee & Shepard.

Specialist

Albrecht, H., and S.C. Dunnett. 1971. *Chimpanzees in Western Africa.* Munich: R. Piper & Co.

Annett, M. 1985. *Left, Right, Hand and Brain: The Right Shift Theory.* Cambridge, Mass: Lawrence Erlbaum & Associates.

Barron, J. N. 1970. The structure and function of the skin of the hand. *British Journal of Surgery of the Hand* 2:93–99.

Beard, K. C., M. F. Teaford, and A. Walker. 1986. New wrist bones of *Proconsul africanus* and *P. nyanzae* from Rusinga Island, Kenya. *Folia Primatologica* 47:97–118.

Beaton, A. 1986. *Left Side, Right Side: A Review of Laterality Research.* New Haven: Yale University Press.

Beck, B. B. 1980. *Animal Tool Behavior.* New York: Garland STPM Press.

Binford, L. R. 1987. Searching for camps and missing the evidence: another look at the Lower Paleolithic. In *The Pleistocene Old World: Regional Perspectives*, O. Soffer, ed., pp. 19–31. New York: Plenum.

Boesch, C. 1991. Handedness in wild chimpanzees. *International Journal of Primatology* 12:541–58.

Boesch, C., and H. Boesch. 1990. Tool use and tool making in wild chimpanzees. *Folia Primatologica* 54:86–99.

Brain, C. K. 1981. *The Hunters or the Hunted? An Introduction to African Cave Taphonomy.* Chicago: University of Chicago Press.

———. 1988. New information from the Swartkrans Cave of relevance to "robust" australopithecines. In *Evolutionary History of the "Robust" Australopithecines.*, F. E. Grine, ed., pp. 311–16. New York: Aldine de Gruyter.

Breuil, H. 1952. *Quatre cents siècles d'art pariétal: les cavernes ornées de l'Age du Renne.* Moñtignac, Dordogne, France: Centre d'Etudes et de Documentation Préhistorique.

Brooker, D. R. 1977. Henry Faulds (1843–1930). *Fingerprint Whorld* 2:59–64.

Bulwer, John. 1974. *Chirologia: or, The natural language of the hand*, and *Chironomia: or, The art of manual rhetoric.* J. W. Cleary, ed. Carbondale: Southern Illinois University Press.

———. 1975. *Chirologia: or, The naturall language of the hand, composed of the speaking motions, and discoursing gestures thereof, whereunto is added Chironomia: or, The art of manual rhetoricke, consisting of the naturall expressions, digested by art in the hand, as the chiefest instrument of eloquence, by historicall manifesto's, exemplified out of the authentique registers of common life, and civill conversation, with types, or chyrograms, a long-wish'd for illustration of this argument*. (Reprint of 1644 ed.). New York: AMS Press.

Bunnell, S. 1944. *Surgery of the Hand*. Philadelphia: J. B. Lippincott.

Bush, M. E., C. O. Lovejoy, D. C. Johanson, and Y. Coppens. 1982. Hominid carpal, metacarpal, and phalangeal bones recovered from the Hadar Formation: 1974–1977 collections. *American Journal of Physical Anthropology* 57:651–77.

Capener, N. 1952. Leonardo's left hand. *The Lancet* 19:813–16.

Cartmill, M. 1972. Arboreal adaptations and the origin of the order Primates. In *The Functional and Evolutionary Biology of Primates*, R. H. Tuttle, ed., pp. 97–122. Chicago: Aldine-Atherton.

———. 1974a. Rethinking primate origins. *Science* 184:436–43.

———. 1974b. Pads and claws in arboreal locomotion. In *Primate Locomotion*, F. A. Jenkins, ed. 45–83. New York: Academic Press.

Clark, W. E. Le Gros. 1936. The problem of the claw in primates. *Proceedings of the Zoological Society of London*, pp. 1–24.

———. 1971. *The Antecedents of Man*. Chicago: Quadrangle Books.

Clarkson, P., and A. Pelly. 1962. *The General and Plastic Surgery of the Hand*. Oxford: Blackwell.

Corballis, M. C. 1983. *Human Laterality*. New York: Academic Press.

Crabtree, D. E. 1975. Comments on lithic technology and experimental archaeology. In *Lithic Technology. Making and Using Stone Tools*, E. Swanson, ed., pp. 105–14. The Hague: Mouton.

Cummins, H. 1941. Ancient finger prints in clay. *Scientific Monthly* 7:389–402.

Cummins, H., and C. Midlo. 1961. *Fingerprints, Palms and Soles*. New York: Dover.

Dart, R. A. 1957. *The Osteodontokeratic Culture of Australopithecus prometheus*. Pretoria: *Transval Museum Memoirs*, no. 10.

Darwin, C. 1871. *The Descent of Man and Selection in Relation to Sex*. London: Murray.

Davis, D. D. 1964. *The Giant Panda. Fieldiana: Zoological Memoirs*, no. 3. Chicago: Field Museum of Natural History.

Day, M. H. 1976a. Hominid postcranial material from Bed I, Olduvai Gorge. In *Human Origins*, G. L. Isaac and E. R. McCown, eds., pp. 363–74. Menlo Park, Calif.: W. A. Benjamin.

———. 1976b. Hominid postcranial remains from the East Rudolf succession. A review. In *Earliest Man and Environments in the Lake Rudolf Basin*, Y. Coppens, F. C. Howell, G. L. Isaac, and R.E.F. Leakey, eds., pp. 507–21. Chicago: University of Chicago Press.

———. 1986. *A Guide to Fossil Man*. 4th edition. Chicago: University of Chicago Press.

Deneberg, V. 1981. Hemispheric laterality and the effects of early experience. *Behavioral and Brain Sciences* 1:1–50.

Efrom, D., and J. P. Foley. 1941. *Gesture and Environment*. New York: King's Crown Press.

Fagot, J., and J. Vauclair. 1991. Prospects for a theory and methodology to study manual lateralization in nonhuman primates. *Psychological Bulletin* 109:76–89.

Falk, D. 1987. Brain lateralization in primates and its evolution in hominids. *Yearbook of Physical Anthropology* 30:107–25.

Finch, G. 1941. Chimpanzee handedness. *Science* 94:117–18.

Galdikas, B.M.F. 1982. Orang-utan tool-use at Tanjung Puting Reserve, Central Indonesian Borneo (Kalimantan Tengah). *Journal of Human Evolution* 11:19–33.

Goodall, J. (see also Van Lawick-Goodall) 1986. *The Chimpanzees of Gombe*. Cambridge, Mass.: Harvard University Press.

Grand, T. I. 1972. A mechanical interpretation of terminal branch feeding. *Journal of Mammalogy* 53:198–201.

Harrison, R. J., and W. Montagne. 1969. *Man*. New York: Appleton-Century-Crofts.

Hockett, C. F., and S. A. Altmann. 1968. A note on design features. In *Animal Communication*, T. A. Sebeok, ed., pp. 61–72. Bloomington, Ind.: Indiana University Press.

Holt, S. B. 1968. *Genetics of Dermal Ridges*. Springfield, Ill.: Charles C. Thomas.

Hopkins, W. D., and R. D. Morris. 1993. Handedness in great apes: a comparative perspective. *International Journal of Primatology* 14:1–26.

Johanson, D. C., C. O. Lovejoy, W. H. Kimbel, T. D. White, S. C. Ward, M. E. Bush, B. M. Latimer, and Y. Coppens. 1982. Morphology of the Pliocene partial skeleton (A.L. 288-1) from the Hadar Formation, Ethiopia. *American Journal of Physical Anthropology* 57:403–51.

Jones, F. W. 1916. *Arboreal Man*. London: Edward Arnold.

———. 1942. *Principles of Anatomy as Seen in the Hand*, 2d ed. London: Belliere, Tindall & Cox.

Jouffroy, F. K. and J. Lessertisseur. 1959. La main des Lemuriens malgache comparée a celle des autres Primates. *Memoires de l'Institut Scientifique de Madagascar* 13A: 195–219.

Kaplan, E. B. 1953. *Functional and Surgical Anatomy of the Hand*. Philadelphia: J. B. Lippincott.

Kidd, W. 1907. *The Sense of Touch in Mammals and Birds*. London: Adam and Charles Black.

Klein, R. G. 1989. *The Human Career*. Chicago: University of Chicago Press.

Kubota, K. 1990. Preferred hand use in the Japanese macaque troop, Arashiyama-R, during visually guided reaching for food pellets. *Primates* 31:393–406.

Landsmeer, J.M.F. 1955. Anatomical and functional investigations of the articulation of the human fingers *Acta anatomica* (Suppl. 24) 25:1–69.

Landsmeer, J.M.F., and C. Long. 1965. The mechanism of finger control, based on electromyograms and location analysis. *Acta anatomica* 60:330–47.

Leakey, L.S.B. 1960. Recent discoveries at Olduvai Gorge. *Nature* 188:1050–52.

Leakey, L.S.B., P. V. Tobias, and J. R. Napier. 1964. A new species of genus *Homo* from Olduvai Gorge. *Nature* 202:7–9.

Leakey, M. D. 1971. *Olduvai Gorge*, Vol. 3, *Excavations in Beds I & II, 1960–1963*. Cambridge: Cambridge University Press.

Lever, J. D. 1963. *The Hand of Man*. Cardiff: University of Wales Press.

Lewis, O. J. 1977. Joint remodelling and the evolution of the human hand. *Journal of Anatomy* 123:157–201.

———. 1989. *Functional Morphology of the Evolving Hand and Foot*. Oxford: Clarendon Press.

Long, C., P. W. Conrad, E. A. Hall, and S. L. Furler. 1970. Intrinsic-extrinsic muscle control of the hand in power grip and precision handling. *Journal of Bone & Joint Surgery* 52-A:853–67.

MacNeilage, P. F. 1991. Patterns of handedness across the primate order. In *Primatology Today*, A. Ehara, T. Kimura, O. Takenaka, and M. Iwamoto, eds., pp. 275–78. Amsterdam: Elsevier.

MacNeilage, P. F., M. G. Studdert-Kennedy, and B. Lindblom. 1987. Primate handedness reconsidered. *Behavioral and Brain Sciences* 10:247–303.

Martin, R. D. 1990. *Primate Origins and Evolution*. Princeton, N.J.: Princeton University Press.

Marzke, M. W. 1971. Origin of the human hand. *American Journal of Physical Anthropology* 34:61–84.

———. 1983. Joint functions and grips of the *Australopithecus afarensis* hand, with special reference to the region of the capitate. *Journal of Human Evolution* 12:197–211.

———. 1986. Tool use and the evolution of hominid hands and bipedality. In *Primate Evolution*, J. G. Else and P. C. Lee, eds., pp. 201–9. Cambridge: Cambridge University Press.

Marzke, M. W., and R. F. Marzke. 1987. The third metacarpal styloid process in humans: origin and functions. *American Journal of Physical Anthropology* 73:415–31.

Marzke, M. W., and M. S. Schakley. 1986. Hominid hand use in the Pliocene and Pleistocene: evidence from experimental archaeology and comparative morphology. *Journal of Human Evolution* 15:439–60.

Marzke, M. W., and K. Wullstein. 1991. Morphological correlates of the power grip in humans. *American Journal of Physical Anthropology*, Supplement 12:126.

Masquelet, A. C., J. Salama, G. Outrequin, M. Serrault, and J. P. Chevrel. 1986. Morphology and functional anatomy of the first dorsal interosseous muscle of the hand. *Surgical and Radiologic Anatomy* 8:19–28.

Mellars, P., and C. Stringer. 1989. *The Human Revolution*. Princeton, N.J.: Princeton University Press.

Musgrave, J. H. 1971. How dextrous was Neanderthal man? *Nature* 233:538–41.

———. 1973. The phalanges of Neanderthal and Upper Palaeolithic hands. In *Human Evolution*, M. H. Day, ed., pp. 59–85. London: Taylor & Francis.

———. 1977. The Neanderthals from Krapina, northern Yugoslavia: an inventory of the hand bones. *Zeitschrift für Morphologie und Anthropologie* 68:150–71.

Napier, J. R. 1952. The attachments and function of the abductor pollicis brevis. *Journal of Anatomy* 86:362–69.

———. 1955. The form and function of the carpo-metacarpal joint of the thumb. *Journal of Anatomy* 89:355–41.

———. 1956. The prehensile movements of the human hand. *Journal of Bone & Joint Surgery* 38B:902–13.

———. 1959. Fossil metacarpals from Swartkrans. *Fossil Mammals of Africa*, no. 17. London: British Museum (Natural History).

———. 1960. Studies of the hands of living primates. *Proceedings of the Zoological Society, London* 134:647–57.

———. 1961a. Prehensility and opposability in the hands of primates. *Symposia of the Zoological Society, London*, no. 5, 115–32.

———. 1961b. Hands and handles. *New Scientist* 9:797–99.

———. 1962. Fossil hand bones from Olduvai Gorge. *Nature* 196:409–11.

———. 1963. The locomotor functions of hominids. In *Classification and Human Evolution*, S. L. Washburn, ed., pp. 178–89. New York: Wenner-Gren Foundation for Anthropological Research.

———. 1964. Functional aspects of the anatomy of the hand. In *Clinical Surgery*, C. Rob and R. Smith, eds., pp. 1–31. Washington: Butterworths.

———. 1965. Evolution of the human hand. *Proceedings of the Royal Institution of Great Britain* 40:544–57.

———. 1976. *The Human Hand.* Carolina Biological Readers, no. 61. Burlington, N.C.: Carolina Biological Supply Co.

Napier, J. R., and P. R. Davis. 1959. The forelimb skeleton and associated remains of *Proconsul africanus*. *Fossil Mammals of Africa*, no. 16. London: British Museum (Natural History).

Nishida, T. 1990. *The Chimpanzees of the Mahale Mountains*. Tokyo: University of Tokyo Press.

Nishida, T., and M. Hiraiwa. 1982. Natural history of tool-using behavior by wild chimpanzees feeding upon wood-boring ants. *Journal of Human Evolution* 11:73–99.

Oakley, K. P. 1958. Tools makyth man. *Smithsonian Report for 1958*, pp. 431–55.

Passingham, R. 1982. *The Human Primate*. San Francisco: W. H. Freeman & Co.

Pickford, M. 1986a. The geochronology of Miocene higher primate faunas of East Africa. In *Primate Evolution*, J. G. Else and P. C. Lee, eds., pp. 19–33. Cambridge: Cambridge University Press.

———. 1986b. Cainozoic paleontological sites of western Kenya. *Münchner Geowissenschaftliche Abhandlungen*, Reihe A, 8:1–151.

———. 1986c. Geochronology of the Hominoidea: a summary. In *Primate Evolution*, J. G. Else and P. C. Lee, eds., pp. 123–28. Cambridge: Cambridge University Press.

———. 1986d. Sexual dimorphism in *Proconsul*. *Human Evolution* 1:111–48.

Posnansky, M. 1959. Some functional considerations on the handaxe. *Man* 59:42–44.

Potts, R. B. 1984. Home bases and early hominids. *American Scientist* 72:338–47.

Rose, M. D. 1984. Hominoid postcranial specimens from the Middle Miocene Chinji Formation, Pakistan. *Journal of Human Evolution* 13:503–16.

Sarasin, P. 1918. Über Rechts- und Linkshändigkeit in der Prähistorie, und die Rechtshändigkeit in der historischen Zeit. *Verhandlungen der Naturforschenden Gesellschaft in Basel* 29:122–96.

Shrewsbury, M. M., and R. K. Johnson. 1983. Form, function and evolution of the distal phalanx. *Journal of Hand Surgery* 8:475–79.

Smith, F. H., and F. Spencer. 1984. *The Origins of Modern Humans.* New York: Liss.

Susman, R. L. 1988a. Hand of *Paranthropus robustus* from Member 1, Swartkrans: fossil evidence for tool behavior. *Science* 240:781–84.

———. 1988b. New postcranial remains from Swartkrans and their bearing on the functional morphology and behavior of *Paranthropus robustus.* In *Evolutionary History of the "Robust" Australopithecines,* F. E. Grine, ed., pp. 149–72. New York: Aldine de Gruyter.

Susman, R. L., and N. Creel. 1979. Functional and morphological affinities of the subadult hand (O. H. 7) from Olduvai Gorge. *American Journal of Physical Anthropology* 51:311–32.

Tokuda, K. On the handedness of Japanese monkeys. *Primates* 10:41–46.

Toth, N. 1985. Archeological evidence for preferential right-handedness in the lower and middle Pleistocene, and its possible implications. *Journal of Human Evolution* 14:607–14.

Trinkaus, E. 1983. *The Shanidar Neandertals.* New York: Academic Press.

———. 1986. The Neandertals and modern human origins. *Annual Review of Anthropology* 15:193–218.

———. 1989a. *The Emergence of Modern Humans.* Cambridge: Cambridge University Press.

———. 1989b. The Upper Pleistocene transition. In *The Emergence of Modern Humans,* E. Trinkaus, ed., pp. 42–66. Cambridge: Cambridge University Press.

———. 1989c. Olduvai Hominid 7 trapezial metacarpal 1 articular morphology: contrasts with recent humans. *American Journal of Physical Anthropology* 80:411–16.

Trinkaus, E., and J. C. Long. 1990. Species attribution of the Swartkrans Member 1 first metacarpals: SK 84 and SKLX 5020. *American Journal of Physical Anthropology* 83:419–24.

Trinkaus, E., and I. Villemeur. 1991. Mechanical advantages of the Neanderthal thumb in flexion: a test of an hypothesis. *American Journal of Physical Anthropology* 84:249–60.

Tuttle, R. H. 1967. Knuckle-walking and the evolution of hominoid hands. *American Journal of Physical Anthropology* 26:171–206.

———. 1969. Knuckle-walking and the problem of human origins. *Science* 166:953–61.

———. 1970. Postural, propulsive, and prehensile capabilities in the cheiridia of chimpanzees and other great apes. *The Chimpanzee,* vol. 2, G. H. Bourne, ed., pp. 167–253. Basel: Karger.

———. 1972. Functional and evolutionary biology of hylobatid hands and feet. In *Gibbon and Siamang,* vol. 1, D. M. Rumbaugh, ed., pp. 136–206. Basel: Karger.

———. 1974. Darwin's apes, dental apes and the descent of man. *Current Anthropology* 15:389–426.

———. 1975. Parallelism, brachiation and hominoid phylogeny. In *Phylogeny of the Primates: A Multidisciplinary Approach*, W. P. Luckett and F. S. Szalay, eds., pp. 447–80. New York: Plenum.

———. 1977. Naturalistic positional behavior of apes and models of hominid evolution, 1929–1976. In *Progress in Ape Research*, G. H. Bourne, ed., pp. 277–96. New York: Academic Press.

———. 1981. Evolution of hominid bipedalism and prehensile capabilities. *Philosophical Transactions of the Royal Society, London* B-292:89–94.

———. 1988. What's new in African paleoanthropology? *Annual Review of Anthropology* 17:391–426.

———. 1992. Hands from newt to Napier. In *Topics in Primatology*, vol. 3: *Evolutionary Biology, Reproductive Endocrinology, and Virology*, S. Matano, R. H. Tuttle, H. Ishida, and M. Goodman, eds., pp. 3–20. Tokyo: University of Tokyo Press.

Tuttle, R. H., and G. W. Cortright. 1988. Positional behavior, adaptive complexes, and evolution. In *Orang-utan Biology*, J. H. Schwartz, ed., pp. 311–30. Oxford: Oxford University Press.

Tuttle, R. H., and D. P. Watts. 1985. The positional behavior and adaptive complexes of *Pan gorilla*. In *Primate Morphophysiology, Locomotor Analyses and Human Bipedalism*, by S. Kondo, ed., pp. 261–88. Tokyo: University of Tokyo Press.

van Lawick-Goodall, J. 1968. Behaviour of free-living chimpanzees in the Gombe Stream Reserve. *Journal of Animal Behaviour* 1:161–311.

Walker, A. C., and M. Pickford. 1983. New postcranial fossils of *Proconsul africanus* and *Proconsul nyanzae*. In *New Interpretations of Ape and Human Ancestry*, R. L. Ciochon and R. S. Corruccini, eds., pp. 325–51. New York: Plenum.

Walker, A., and M. Teaford, 1989. The hunt for *Proconsul*. *Scientific American* 260:76–82.

Walker, A., M. F. Teaford, and R. E. Leakey. 1986. New information concerning the R114 *Proconsul* site, Rusinga Island, Kenya. In *Primate Evolution*, J. G. Else and P. C. Lee, eds., pp. 143–49. Cambridge: Cambridge University Press.

Warren, J. M. 1953. Handedness in the rhesus monkey. *Science* 118:622–23.

———. 1980. Handedness and laterality in humans and other animals. *Physiological Psychology* 8:351–59.

Wynn, T., and W. C. McGrew. 1989. An apes's view of the Oldowan. *Man* 24:383–98.

Index

The Princeton Science Library

Edwin Abbott Abbott	**Flatland: A Romance of Many Dimensions** With a new introduction by Thomas Banchoff
Friedrich G. Barth	**Insects and Flowers: The Biology of a Partnership** Updated by the author
Marston Bates	**The Nature of Natural History** With a new introduction by Henry Horn
John Bonner	**The Evolution of Culture in Animals**
A. J. Cain	**Animal Species and Their Evolution** With a new afterword by the author
Paul Colinvaux	**Why Big Fierce Animals Are Rare**
Peter J. Collings	**Liquid Crystals: Nature's Delicate Phase of Matter**
Pierre Duhem	**The Aim and Structure of Physical Theory** With a new introduction by Jules Vuillemin
Manfred Eigen & Ruthild Winkler	**Laws of the Game: How the Principles of Nature Govern Chance**
Albert Einstein	**The Meaning of Relativity** Fifth edition
Niles Eldredge	**Time Frames: The Evolution of Punctuated Equilibria**
Richard P. Feynman	**QED: The Strange Theory of Light**
J. E. Gordon	**The New Science of Strong Materials, or Why You Don't Fall through the Floor**
Richard L. Gregory	**Eye and Brain: The Psychology of Seeing** Revised, with a new introduction by the author
J.B.S. Haldane	**The Causes of Evolution** With a new preface and afterword by Egbert G. Leigh
Werner Heisenberg	**Encounters with Einstein, and Other Essays on People, Places, and Particles**
François Jacob	**The Logic of Life: A History of Heredity**

Rudolf Kippenhahn	**100 Billion Suns: The Birth, Life, and Death of the Stars** With a new afterword by the author
Hans Lauwerier	**Fractals: Endlessly Repeated Geometrical Figures**
John Napier	**Hands** Revised by Russell H. Tuttle
J. Robert Oppenheimer	**Atom and Void: Essays on Science and Community** With a preface by Freeman J. Dyson
John Polkinghorne	**The Quantum World**
G. Polya	**How to Solve It: A New Aspect of Mathematical Method**
Hazel Rossotti	**Colour, or Why the World Isn't Grey**
David Ruelle	**Chance and Chaos**
Henry Stommel	**A View of the Sea: A Discussion between a Chief Engineer and an Oceanographer about the Machinery of the Ocean Circulation**
Hermann Weyl	**Symmetry**